# Evolving Brains

John Morgan Allman

SCIENTIFIC
AMERICAN
LIBRARY

A division of HPHLP
New York

*Cover and text design: Victoria Tomaselli*
*Illustration: Joyce A. Powzyk and Fine Line Studio*

Library of Congress Cataloging-in-Publication
Allman, John Morgan.
  Evolving brains / John Morgan Allman.
  p. cm. — (Scientific American Library, 1040–3213; no. 68)
  Includes bibliographical references and index.
  ISBN 0-7167-5076-7
  1. Brain—Evolution.  I. Title.  II. Series: Scientific American
QP376.A423 1999
573.8'6—dc21
                                        98-37576
                                        CIP

ISSN 1040-3213

Printed in the United States of America
First paperback printing, 2000

Scientific American Library
A division of HPHLP
New York

Distributed by W. H. Freeman and Company
41 Madison Avenue, New York, NY 10010
Houndmills, Basingstoke RG21 6XS, England

University of
Chester

To my father,
who encouraged me to ask questions

# CONTENTS

# PREFACE

On a crisp October evening in 1966, I decided to study brain evolution. I wanted to discover the underlying physiological mechanisms responsible for brain evolution by comparing living animals. I knew that primates excelled in their use of vision, and I thought a particularly useful approach would be to trace how the visual image on the retina was transformed into neural signals in the brains of primates. I was a graduate student in anthropology, and I knew that I would have to become proficient in neurophysiology to undertake this investigation. Clark Howell, my adviser at the University of Chicago, enthusiastically supported my rather unconventional goal and suggested that I conduct my research at the Laboratory of Neurophysiology at the University of Wisconsin while still a graduate student at Chicago. Anthropologists typically go off to some exotic locality to conduct their thesis research, so my trip was shorter than most. I was doubly fortunate that Clinton Woolsey and Wally Welker welcomed me to the Laboratory.

At Wisconsin I began a fruitful collaboration with Jon Kaas that led to our early mapping studies of the visual cortex in owl monkeys. Our studies were based on the pioneering microelectrode mapping techniques developed by Wally and Vicente Montero. Jon and I uncovered a far larger and more complex set of areas than anyone had expected, and the mapping and study of the functions of these areas continue to this day in monkeys and humans in many laboratories throughout the world. In 1974, I moved to Caltech, which proved to be a wonderful and nurturing scientific home, where I continued my studies of the functional organization of visual cortex but remained guided by the broader questions of how brains have evolved. Much of this work appears in different parts of this book, and it would not have been possible without my students and postdoctoral fellows: Jim Baker, Zachary Berger, Leslie Brothers, Allan Dobbins, Emmanuel Gilissen, Atiya Hakeem, Andrea Hasenstaub, Richard Jeo, Roshan Kumar, Jason Lee, Colin MacDonald, Sanjoy Mahajan, Paul Manis, EveLynn McGuinness, Todd McLaughlin, Bassem Mora, Bill Newsome, Steve Petersen, Kathy Rockland, David Rosenbluth, Aaron Rosin, Stanzi Royden, Gisela Sandoval, Rahul Sarpeshkar,

Marty Sereno, and Dave Sivertsen. Francis Miezin provided enormous assistance in the early years when computers were less than user friendly. More recently Laura Rodriguez and Holli Weld have been a great help in keeping the laboratory running smoothly.

The greatest satisfactions of a life in science are the people with whom one shares ideas and stimulating discussions. I am grateful to Ralph Adolphs, Leslie Aiello, Karen Allendoerfer, Steve Allman, Richard Andersen, Jane Arnault, Mary and Howard Berg, Richard Bing, Roger Bingham, Jim Bower, Bill Brownell, Ted Bullock, Chris Burbeck, Boyd Campbell, Matt Cartmill, Deborah Castleman, Sheila and David Crewther, Francis Crick, Susan Crutchfield, Antonio and Hannah Damasio, Bob Desimone, Ursula Dräger, Caleb Finch, Barbara Finlay, Scott Fraser, John Gerhart, Charles Gilbert, Don Glaser, Richard Gregory, David Grether, Charles Gross, Patrick Hof, David Hubel, Tom Insel, Russ Jacobs, Jukka Jernvall, Bela Julesz, Jon Kaas, Bettyann Kevles, Joe Kirschvink, Christof Koch, Arianne Faber Kolb, Mark Konishi, Leah Krubitzer, Gilles Laurent, Ed Lewis, Margaret Livingstone, Bob Martin, John Maunsell, Carver Mead, Mike Merzenich, Eliot Meyerowitz, John Montgomery, Ken Nakayama, Pietro Perona, Jack Pettigrew, Charles Plott, Pasko Rakic, Michael Raleigh, V. S. Ramachandran, Brian Rasnow, John Rubenstein, Ken Sanderson, Arnold Scheibel, Terry Sejnowski, Roger Sperry, Larry Squire, Steven Suomi, Richard Taylor, Margaret and Johannes Tigges, Roger Tootell, Carol Travis, Leslie Ungerleider, David Van Essen, Wally Welker, Torsten Wiesel, Rob Williams, Margaret Wong-Riley, Patricia Wright, and Steve Zucker.

Jonathan Cobb, then at W. H. Freeman and Company, gave me the opportunity to write this book, and he and Susan Moran provided valuable feedback during the early stages of the writing. Philip McCaffrey provided enthusiastic support when Jonathan moved to another publishing house. I am especially grateful to Nancy Brooks and Georgia Lee Hadler, whose continual good cheer and skillful editing have made the writing of this book a joyful experience. I am also very grateful to Joyce Powzyk for creating much of the original artwork; Joyce possesses the rare combination of great artistic sensitivity and a strong scientific knowlege of brains and behavior. Leslie Wolcott drew the richly expressive monkey faces depicted in Chapter 6. Larry Marcus tirelessly sought out many of the photographs. Howard Berg, Patrick Hof, Jon Kaas, Ed Lewis, Robert Martin, Marty Sereno, and Wally Welker generously provided illustrations from their work. Shady Peyvan and Tess Legaspi obtained countless books and papers that I needed from libraries throughout the country. Wayne Waller digitized several of the images used in this book. Howard Berg, Roger Bingham, Deborah Castleman, Tony Damasio, Caleb Finch, Jukka Jernvall, Jon Kaas, Bettyann Kevles, Michael Raleigh, Terry Sejnowski, and Holli Weld have kindly read the manuscript and offered many helpful suggestions. My companion and friend from childhood, Carolyn Robinson Nesselroth, was a wonderful sounding board for the development of the ideas expressed in this book. Science and scientific communication, like the evolution of the human brain itself, depend profoundly on social networks of cooperative support.

New information contributing to our understanding of brain evolution is appearing at an accelerating rate. This summer, as I put the finishing touches to *Evolving Brains*, a flood of important new findings were reported that have clarified significant issues; these have been incorporated into the book. Our knowledge of brain evolution is itself very rapidly evolving.

It occurs to me as I finish this book that words derived from the verb "vary" appear frequently throughout the text. This is because, as Darwin and Wallace recognized long ago, natural variation provides the raw material for evolution, and the analysis of how features vary in organisms is fundamental to understanding evolution. These words also arise frequently in another context related to changes in the environment. Brains are one of several means that animals use to buffer themselves from environmental variations that would otherwise threaten their existence. The study of variation is the key to understanding how brains have evolved and even to life itself.

Pasadena, California
August 1998

# EVOLVING BRAINS

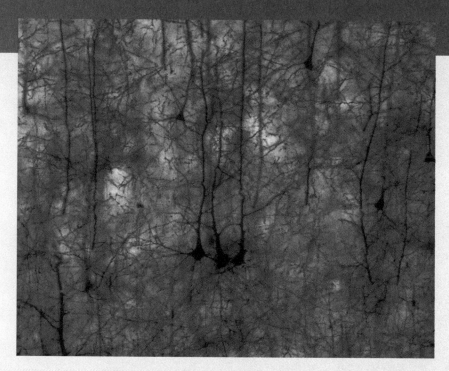

For their size, brains are the most complex systems known. This photomicrograph illustrates the myriad branching patterns of dendrites of neurons in the mammalian neocortex. Only a small fraction of the total number of neurons present is stained; the actual dendritic pattern is vastly more complex.

# Brain Basics

All living things have much in common, in their
chemical composition, their germinal vesicles, their
cellular structure, and their laws of growth and
reproduction. Therefore I should infer that probably
all the organic beings which have ever lived on the
earth have descended from some one primordial form.

Charles Darwin,
*The Origin of Species,* 1859

Given that all organisms share a common ancestry, why is it that they differ so greatly in their capacities to sense, remember, and respond to the world about them? How did we gain our ability to think and to feel? How do we differ from other organisms in these capacities? Our brain endows us with the faculties and the drive to ask these fundamental questions. The answers depend crucially on understanding how brains have evolved. This inquiry into brain evolution is interdisciplinary and multifaceted, based on converging evidence obtained from the study of the genetic regulation of development, the geological history of the earth, and the behavioral ecology of animals, as well as from direct anatomical and physiological studies of brains of animals of different species. From this investigation three themes will emerge: that the essential role of brains is to serve as a buffer against environmental variation; that every evolutionary advance in the nervous system has a cost; and that the development of the brain to the level of complexity we enjoy—and that makes our lives so rich—depended on the establishment of the human family as a social and reproductive unit.

I will begin by considering one of the basic problems faced by all organisms: how to find food and avoid hazards in a constantly changing world. This leads to the question of how nervous systems detect and integrate the vast array of information available to them and derive from this flood of data adaptive behavioral responses. The evolution of nervous systems depended on a unique mechanism for communication, the action potential, a self-renewing electrical signal that moves along specialized neural fibers called axons that serve as the wires connecting nerve cells. By permitting the development of large nervous systems, this mechanism for neuronal communication made possible the emergence of complex and diverse forms of animal life.

## Why Brains?

Brains exist because the distribution of resources necessary for survival and the hazards that threaten survival vary in space and time. There would be little need for a nervous system in an immobile organism or an organism that lived in regular and predictable surroundings. In the chaotic natural world, the distribution and location of resources and hazards become more difficult to predict for larger spaces and longer spans of time. Brains are informed by the senses about the presence of resources and hazards; they evaluate and store this input and generate adaptive responses executed by the muscles. Thus, when the required resources are rare, when the distribution of these resources is highly variable, when the organism has high energy requirements that must be continuously sustained, and when the organism must survive for a long period of time to reproduce, brains are usually large and complex. In the broadest sense then, brains are buffers against environmental variability. This theme will emerge time and again as we explore how brains have evolved.

Some of the most basic features of brains can be found in bacteria because even the simplest motile organisms must solve the problem of locating resources and avoiding toxins in a variable environment. Strictly speaking, these unicellular organisms do not have nervous systems, but nevertheless they exhibit remarkably complex behavior. They sense their environment through a large number of receptors and store this elaborate sensory input in the form of brief memory traces. Moreover, they integrate the inputs from these multiple sensory channels to produce adaptive movements. The revolution in our understanding of genetic mechanisms has made it possible to determine how these brainlike processes work at a molecular level in bacteria.

## Brainlike Functions in Unicellular Organisms

We have a particularly intimate relationship with the coliform bacteria, *Escherichia coli*, because they inhabit our intestinal tract and aid our digestion (although occasionally they can become less than congenial dining companions). The mechanisms whereby they sense, remember, and move about their environment provide an excellent model for the basic features of nervous systems, albeit in an organism contained within a single cell and lacking a brain in a conventional sense. The discovery by Julius Adler, Daniel Koshland, Melvin Simon, and others of a large number of genetic mutations that selectively influence the underlying chemistry of bacterial life has made it possible to dissect the functional organization of the systems for sensory responses, memory, and motility in exquisite detail, and *E. coli* has been a favorite subject of these investigations.

The bacterium senses its environment through its receptors, which are protein molecules embedded in the cell wall that bind to

## Genetic Mechanisms

Genetic mechanisms are basic to the consideration of all aspects of evolution. In 1953, James Watson and Francis Crick discovered the structural basis for the storage of genetic information in the chromosomes. They found that the information-carrying molecule is deoxyribonucleic acid, DNA, which consists of two matched helices that are connected by base pairs made of four different compounds. Adenosine (A) pairs with thymidine (T), and cytidine (C) pairs with guanine (G). The sequence of base pairs is the genetic code, which is nearly universal in all living organisms. The code is read from one direction in one strand. Three-letter sequences, triplets, specify each amino acid, and the sequence of triplets in turn specifies the chain of amino acids that makes up a protein. There are $4^3$, or 64, possible triplet sequences. Each of 61 triplets encodes for one of the 20 amino acids. Thus some amino acids are specified by more than one triplet, although no triplet specifies more than one amino acid. The other three triplets are stop codons that signal the end of a particular protein. The complete sequence of triplets that encodes a protein is a gene.

Homologous genes have similar DNA sequences and are derived from a replication of a common ancestral gene. By comparing the differences in the sequences of homologous genes, it is possible to make a rough estimate of the time elapsed since the original divergence from the common ancestral gene. Adjacent to the genes are promoters, DNA sequences that modify the activity of the gene. The bulk of the DNA is neither genes nor promoters, and this material has sometimes been termed "junk" DNA because it has no known function. Much of the junk DNA consists of genes that have been rendered inactive through mutations. Junk DNA may be an excellent place to dig for fossils. By providing information about genes that were active in the past, junk DNA may prove to be an important source for reconstructing evolutionary history.

In addition to the DNA in the chromosomes, which are in the nucleus of the cell, there is also DNA in bodies called mitochondria, which are located in the cellular fluid, or cytoplasm. The mitochondria are tiny generators that produce energy for the cell, and their DNA contains part of the sequence that encodes the energy producing enzyme, cytochrome oxidase. Lynn Margulis proposed that the mitochondria are derived from ancient bacteria that long ago infected cells that were ancestral to animals, plants, and fungi. These invading bacteria brought with them the

specific chemicals outside the cell and communicate with mechanisms inside the cell. *E. coli* has more than a dozen different types of receptors on its surface. Some are specialized for the detection of different nutrients, such as particular types of sugar, which provide energy, or amino acids, which are the building blocks of proteins; other receptors are responsive to toxins, such as heavy-metal ions.

mechanisms for producing energy through the oxidation of carbohydrates into carbon dioxide and water, and these mechanisms became indispensable to the host cells and were in part responsible for their evolutionary success. Mitochondrial DNA is the residue of the genes of these ancient bacteria. The mitochondrial DNA present in the egg is transmitted from mother to offspring, and thus is inherited exclusively in the maternal line. Comparisons in the sequences of mitochondrial DNA are very useful for estimating the timing of evolutionary events.

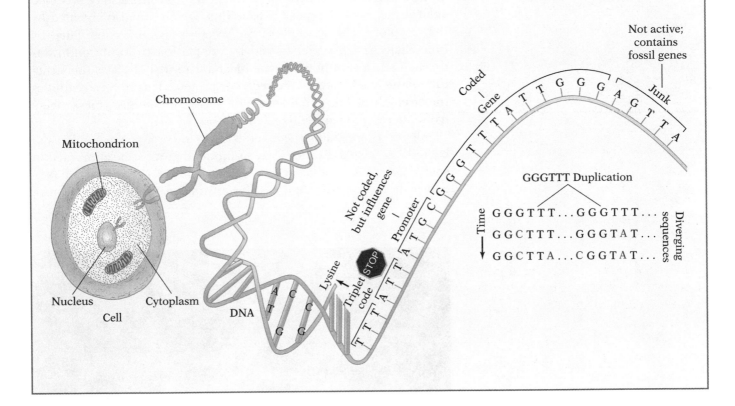

Bacteria possess highly developed sensory systems for the detection of nutrients, energy sources, and toxins, and the capacity to store and evaluate the manifold information provided by these diverse receptors. The final outcome of this sensory integration is the decision to continue swimming in the same direction or tumble into a different course. Thus some of the most fundamental features of brains such as sensory integration, memory, decision-making, and the control of behavior, can all be found in these simple organisms.

Detection of an increasing concentration of a particular nutrient or a decreasing concentration of a toxin causes the bacterium to swim forward propelled by its flagella. The detection of a gradient requires the memory and comparison of previous receptor responses occurring during the past few seconds. The relative input from each type of receptor must also be balanced to adjust for the mix of nutrients available at a particular place in the environment and the needs of the bacterium. The benefits of the nutrients must also be weighed against the risks of exposure to toxins, which are also signaled by the receptors. The strength of the receptor signal is modulated by immediate past experience and by adaptation to local conditions through an elegant biochemical mechanism that changes the structure of the inside loop of the receptor protein. This in turn regulates the strength of the signal sent by a series of chemical messengers from the receptor to the flagellar motors.

All brains, including these very simple integrative mechanisms in bacteria, receive a diverse array of inputs that must be combined in such a way as to produce a very much smaller set of behavioral outcomes. In the case of *E. coli,* the organism can either swim forward,

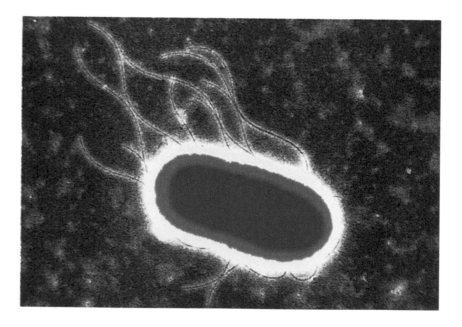

A bacterium, *Escherichia coli.* The surface of the bacterium is bright yellow in this color-enhanced image. The flagella are long strands extending from the bacterium into the surrounding medium. The flagella shown here are in the tumble mode.

Maltose Sugar Gradient

| Low concentration | High concentration | Low concentration | High concentration |
|---|---|---|---|

| Run | Run | Tumble | Run |
|---|---|---|---|

During a run, the six flagella of *E. coli* are gathered together to form a propeller. When the receptors detect a decreasing concentration of the sugar resource, they signal the flagellar motors to reverse, causing the flagella to flail about uselessly in a tumble and the bacterium to change its course. The bacterium then reverses its motors again and begins a run in a different direction. The bacterium is only very briefly in the tumble mode; it moves nearly constantly.

in a movement termed a "run," or reverse the direction of its flagellar motors, stalling its forward motion. The stall causes the organism to go into a "tumble" and thus change its direction. The flagellar motor is one of the most extraordinary engineering achievements in biology. Incredibly, it actually has a crankshaft passing through the cell membrane that is connected to the external propeller, the flagellum. The crankshaft is driven by up to eight ratchets that derive their power from the flow of hydrogen ions.

Some unicellular organisms even possess visual systems that exhibit remarkable similarities to our own. *Halobacterium salinarium*, which lives in salt marshes, derives its energy from light-driven mechanisms that work optimally in orange light. John Spudich has shown that it possesses a light-sensitive pigment that has a similar molecular structure to rhodopsin, the photoreceptive pigment in the eyes of vertebrates. In both *Halobacterium* and vertebrates, the pigment is a chain of amino acids that loops back and forth through the cell membrane seven times and surrounds the compound retinal, which changes its shape in response to light. Rhodopsin in *Halobacterium* is maximally sensitive to orange light, and the receptors work in much the same way as do the nutrient receptors in *E. coli* to cause the bacterium to swim toward sunlight, its energy source.

*Chlamydomonas* is another unicellular organism that possesses a simple eye; its means of propulsion is different, but equally elegant. A free-swimming algae that depends on sunlight to drive photosynthesis, *Chlamydomonas* uses blue-green light to orient its swimming

| Specifications of Flagellar Motor | |
| --- | --- |
| Diameter | 1 micro inch |
| Speed | 6000 rpm |
| Power output | 1/10 micro-micro-micro hp |
| Power per unit weight | 10 hp per pound |
| Power source | proton current |
| Cylinders | 8 |
| Number of different kinds of parts | 30 |
| Gears | 2: forward (run) and reverse (tumble) |

(Courtesy of Howard Berg's Engine Shop.)

with respect to its energy source. It is equipped with two flagella attached to its front end, which instead of rotating like bacterial flagella, bend to perform a breaststroke. The bending is accomplished by the differential sliding of 9 of the 10 pairs of microtubules running the length of each flagellum. (This is the classic structure of cilia, which also perform many functions in multicellular organisms.) The eye spot, located on the cell's equator, also contains a photoreceptive compound very similar to vertebrate rhodopsin. Like the chemotactic response in bacteria, this detection of the direction of the light source requires a short-term memory to enable the comparison of light intensities picked up at different phases of the cell's rotation. However, unlike bacteria, which can only run or tumble, *Chlamydomonas* can change its swimming direction by adjusting the bending movements of its two flagella.

Cilia are a very elegant way to propel unicellular organisms. Because they project into the fluid surrounding the cell, they are also very well positioned to sense the environment. With the advent of multicellular organisms, cilia took on remarkable new functions: for example, in the vertebrate eye the rhodopsin pigment migrated into cilia to become the photoreceptors in the retina. Olfactory and other sensory receptors are also modified cilia.

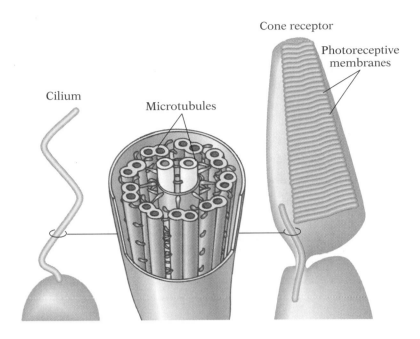

The cilium extends from the cell into the surrounding fluid medium. The pairs of microtubules slide with respect to one another, causing the cilium to move. Cilia are well suited to perform sensory functions as well, and many types of receptors are modified cilia. For example, cone receptors in the vertebrate retina, which are responsible for color vision, are modified cilia in which stacks of photoreceptive membranes have replaced most of the microtubular structure.

## Controlling the Flow of Information

Cells, like people, are immersed in a flood of information that they must evaluate in order to generate an adaptive response. Even *E. coli* must integrate information from more than a dozen different receptor types to make the simple binary decision as to whether to rotate its flagellar motors clockwise or counterclockwise. Thus a tremendous reduction of data takes place as information collected from diverse receptors leads to a very limited set of motor responses. One advantage possessed by multicellular organisms is that they can channel this flood of information by creating a dam between the external world and the interior of the organism, enabling them to control the chemical environment and thus the flow of information within the organism. The chemical environment is regulated by channels located in the membranes of cells that control the passage of specific ions, such as sodium and potassium. Accompanying the compartmentalization of information are specializations of cell function. Cells specialized for receptor function are located on the surface of the organism. Other cells specialized for the transmission and analysis of information are located in the protected interior and are linked to effector cells, usually muscles, which produce adaptive responses.

Human behavior has more in common with bacteria than might be supposed. Floor traders in the stock market respond to a wide variety of inputs concerning resources and hazards that lead them to the binary decison to buy or sell a particular stock. Like bacteria, they typically operate over a short time interval.

Brains are made of neurons, cells specialized for processing information. Like the unicellular organisms, neurons have receptors located on the cell surface for different chemicals including various amino acids, reflecting the ancient nutritive function of these signaling compounds in unicellular organisms. As do unicellular organisms, neurons integrate the diverse array of incoming information from the receptors, which in neurons may result in the firing of an action potential rather than swimming toward a nutrient source as in the unicellular organisms. This integration may also result in an inhibition of the tendency to fire an action potential in a neuron just as it may result in the suppression of the tendency to swim forward in unicellular organisms.

Neurons have a characteristic architecture in which the chemical receptors tend to be located on branching structures, the dendrites, which extend from the cell body. Dendrites with their branches look like trees, and the term is derived from the Greek word for "tree," *dendron*. The dendrites increase the receptive surface of the neurons, but their length is constrained by their electrical properties so that they rarely extend more than a few millimeters from the cell body. The integration and storage of information occur primarily in the dendrites. Neurons require a great deal of energy to maintain the ionic balance between themselves and their surrounding fluids, which is constantly in flux as a result of the

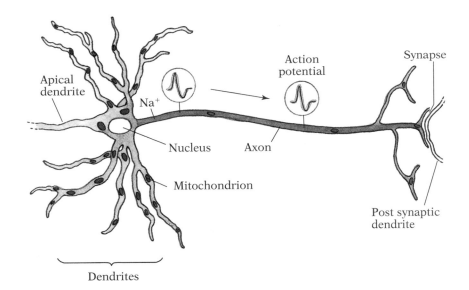

The dendrites vastly increase the surface area of the neuron and contain many mitochondria that generate the energy necessary to maintain ionic gradients across the membrane between the interior of the dendrite and the fluid space outside the dendrite. The action potential is a two-phase electrical pulse that originates where the axon leaves the cell body of the neuron. The action potential travels down the axon to its terminals, where it causes the release of a chemical neurotransmitter at synapses, the sites of connections between neurons. Neurotransmitter is released from the axon terminal into the synaptic cleft and then binds to receptors in the adjacent dendrite.

opening and closing of channels through the neuronal membranes. The fine branches of the dendrites greatly increase the receptive membrane surface area of neurons, and thus the smaller branches contain large concentrations of mitochondria, which generate the energy necessary to maintain this ionic equilibrium. The large energy requirements of nervous systems have constrained their development, which is an important factor influencing the evolution of large brains.

Neurons are dynamically polarized, so that information flows from the fine dendrites into the main dendrites and then to the cell body, where it is converted into all-or-none signals, the action potentials, which are relayed to other neurons by the axon, a long, wirelike structure. The action potential is initiated by the opening of voltage-sensitive sodium channels in the membrane of the axon at the point where the axon emerges from the cell body. Sodium ions rush into the neurons from the extracellular fluid, resulting in a transient change in the voltage difference between the neuron and the surrounding environment. The action potential travels like a wave from the cell body down the axon. Each action potential has approximately the same size, shape, and duration, all of which are maintained as the action potential travels down the axon. The action potential enables the neuron to communicate rapidly with other neurons over sizable distances, sometime more than a meter away.

The axon branches into a series of terminals that form connections with the dendrites of other neurons at sites called synapses. When the action potential reaches an axon terminal, it causes the terminal to secrete a chemical messenger (neurotransmitter), generally an amino acid or its derivative, which binds to receptors in the post-synaptic neurons on the far side of the synaptic cleft.

Nervous systems are like hybrid computers that utilize both analog and digital signals and thus gain the advantages offered by both modes of computation. The strength of analog signals varies across a continuum, whereas digital signals are all or none. The dendrites integrate thousands of synaptic inputs, each of which has a small influence on the voltage within the dendrite in a manner similar to an analog computer. Thus they have the capacity to represent a great deal of information. However, as Rahul Sarpeshkar points out, because analog computations vary across a continuum, they are vulnerable to noise and are prone to drift, which makes them unsuitable for long-distance communication or integration into a large system. By contrast, the all-or-nothing action potentials can represent the integrated output of the dendrites in a discrete manner and relay this information faithfully via the axons to other parts of the system.

Action potentials and voltage-gated sodium channels are present in jellyfish, which are the simplest organisms to possess nervous systems. The communication among neurons via action potentials and its underlying mechanism, the voltage-gated sodium channel, were essential for the evolution of nervous systems, and without nervous systems complex animals could not exist. The development of this basic neuronal mechanism set the stage for the great proliferation of animal life that occurred during the Cambrian period, more than half a billion years ago.

Among the less spectacular of these Cambrian animals were the early chordates, which possessed very simple brains. From this modest beginning evolved the earliest vertebrates, which were small predators with a keen sense of smell and an enhanced capacity to remember odors. Some of these early fish developed a unique way to insulate their axons by wrapping them with a fatty material called myelin, which greatly facilitated axonal transmission and evolution of larger brains. Some of their descendants, which also were small predators, crawled up on the muddy shores and eventually took up permanent residence on dry land. Challenged by the severe temperature changes in the terrestrial environment, some experimented with becoming warm-blooded, and the most successful became the ancestors of birds and mammals. Changes in the brain and parental care were a crucial part of the set of mechanisms that enabled these animals to maintain a constant body temperature. Our ancestors, the early primates, were also small predators, with large frontally directed eyes, grasping hands, and enlarged brains.

Animals with large brains are rare—there are tremendous costs associated with large brains. The brain must compete with other organs in the body for the limited amount of energy available, which is a powerful constraint on the evolution of large brains. Large brains also require a long time to mature, which greatly reduces the rate at which their possessors can reproduce. Because large-brained infants are slow to develop and are dependent on their parents for such a long time, the parents must invest a great deal of effort in raising their infants. The evolution of very large brains requires sustained care for very dependent and slowly developing offspring. The evolution of large brains in humans depended crucially on the establishment of the extended family to provide this care.

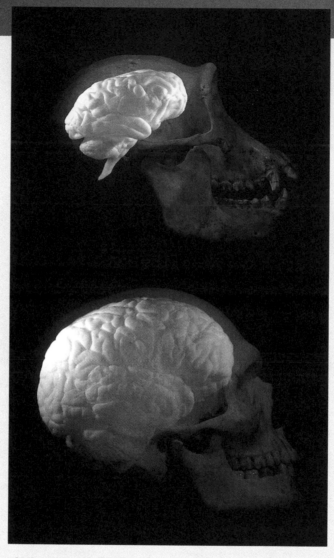

The human and the chimpanzee have similar body sizes, but the human brain is about three times larger than the chimpanzee's.

# Comparing Brains

Always remember that the one true, certain, final,
and all-important difference between you and an
ape is that you have a hippopotamus major in your
brain, and it has none.

Professor Ptthmllnsprts
in Charles Kingsley's *Water Babies,* 1863

Professor Ptthmllnsprts ("put-them-all-in-spirits") was a parody of the eminent Victorian anatomist Richard Owen, who claimed that he had identified a unique structure in the human brain that he called hippocampus minor. Owen had concluded, "Peculiar mental powers are associated with the highest form of brain, and their consequences wonderfully illustrate the value of the cerebral character; according to my estimate of which, I am led to regard the genus *Homo*, as not merely a representative of a distinct order, but of a distinct subclass of the Mammalia, for which I propose the name of ARCHENCEPHALIA." Owen's ruling-brain classification for humans was brought down by Thomas Henry Huxley, who showed that the hippocampus minor was present and well developed in the brains of other primates. This cautionary tale illustrates the risks of concluding that any structure is uniquely human, but nevertheless the comparison of brains provides much useful information. Brains vary greatly in both size and in neural circuitry. Some brain structures are remarkably constant in all vertebrates and presumably perform very basic functions common to all vertebrates. Other structures are extremely variable. Comparing these constant and variable structures in different animals sheds light on the evolutionary history of the brain in vertebrates.

## Weighing Brains and Comparing Structures

In comparing whole brains, the simplest approach is to weigh various animals and their brains. Brain tissue has about the same density in different animals, but unfortunately a given mass of brain tissue cannot be easily related to its functional capacity. This is because no reliable technique exists for measuring the functional capacity of brains across different species. Moreover, brains consist not only of neurons but also of various supportive cells, most notably the glia and blood vessels. The glia serve to guide the migration of neurons in development, regulate the chemical balance of extracellular fluids in the brain, and manufacture myelin, the fatty material that serves to insulate axons and facilitate their transmission of action potentials.

A useful way to compare the weights of the brains of different animals was introduced by Harry Jerison, whose 1973 book, *Evolution of the Brain and Intelligence*, was a pioneering study of brain size. He noted that when the body weights of different animals are plotted against brain weights, the animals of particular groups fall

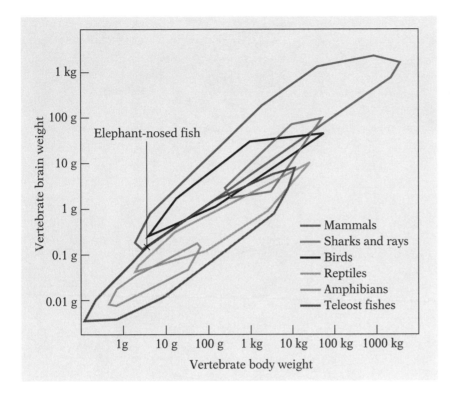

The relationship between brain and body weight in vertebrate groups, plotted in grams on logarithmic scales on which each tick marks a tenfold change. The members of each particular vertebrate group, such as the mammals, fall within a well-defined polygon. The teleosts are a large group of bony fish, distinct from the cartilaginous sharks and rays.

within well-defined polygons. For each group, the polygon rises to the right, indicating that brain weight tends to increase with body weight. Note, however, that the ratio is not constant. In each class of vertebrate, the weight of the brain always increases more slowly than does the weight of the body. The polygons for birds and mammals lie entirely above those for amphibians and reptiles, indicating a marked advance in quantitative brain evolution in the warm-blooded animals.

However, the polygon for fish overlaps the avian and mammalian distributions to some extent, indicating that large brain size evolved independently in fish as well as in warm-blooded vertebrates. This should not be surprising in view of the 25,000 living species of fish, which make them the most diverse class of vertebrates. Several types of fish have relatively large brains for their body size. The mormyrid fish—the elephant-nosed fish is an example—probe their muddy waters for mates and prey with pulses of

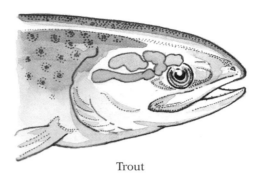

Trout

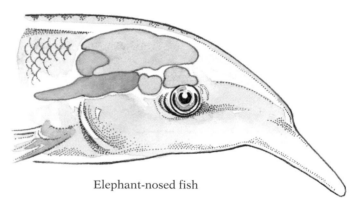

Elephant-nosed fish

The brain of a typical fish, a trout, compared with the brain of the elephant-nosed fish, a mormyrid. The overall brain size, and especially the blue-shaded component, the cerebellum, is much larger in the mormyrid.

electric current, generated by their electric organs, and analyze the reflected currents with special electroreceptors. Their large relative brain size is related to their capacity to discriminate features in their world on the basis of these reflected electric signals. The brains of mormyrids require a great deal of energy to operate. Goran Nilsson, a physiologist at Uppsala University, found that the brain in mormyrids takes 60 percent of the oxygen used by the entire body in these fish. The second group of fish with large brains are the sharks and rays, which are among the most formidable predators in the ocean. Interestingly, the nonpredatory basking shark has the smallest brain for its body size within this group. This is one of many instances in which brain evolution appears to be linked to predatory behavior.

The comparative data for brain weight in different vertebrates support three important conclusions that I shall return to throughout this book. First, in order to compare the brain weights of different-sized animals one must take into account their body weights. Second, the larger brain weights of mammals and birds are associated with their much higher energy requirements. Third, expansion of brain size has occurred in the different classes of vertebrates; it is not unique to mammals.

Another technique is to compare the anatomical organization and circuitry in similar structures in the brains of different vertebrates. Brains are made up of circuits of awesome complexity. As a way into these vast systems, I will compare two brain structures, one at the bottom and the other at the top of the brain. Comparing the two reveals the radically different courses in evolution taken by these structures. The bottom structure, the network of serotonergic neurons in the brain stem, was present in the earliest vertebrates and has retained a remarkably constant anatomical position throughout vertebrate evolution. The serotonergic neurons are so named because they secrete from their axon terminals the neurotransmitter serotonin. The top structure, the neocortex, is much more recently evolved and is extremely variable in its anatomical organization. The term "cortex" refers to the outer shell of the brain, and "neo" implies that it is new. The neocortex is found only in mammals, although it is related to forebrain structures found in other vertebrate classes. The neocortex has expanded enormously in the brains of humans and other advanced mammals.

## The Serotonergic Stabilizer

Serotonin was discovered in 1948 by the biochemist Maurice Rapport and his colleagues, working at the Cleveland Clinic. They found that it caused blood vessels to constrict and derived the name from the combination of the Latin words for "blood," *serum*, and "stretching," *tonus*. Serotonin was soon found by other investigators to cause contractions of the gut. However, subsequent studies found that serotonin could have the *opposite* effects in the blood vessels and gut, indicating that serotonin has a complex modulatory

The serotonergic synapse. Serotonin is synthesized from tryptophan in the presynaptic axon terminal and released into the synaptic cleft. The released serotonin either binds to the postsynaptic receptors in the dendrite of the post-synaptic neuron or is absorbed back into the axon terminal by the transporter reuptake mechanism. Prozac and similar drugs inhibit the reuptake mechanism, thus increasing the concentration of serotonin within the synpatic cleft. Estrogen inhibits the expression of the gene for the serotonin transporter. The basic role of serotonin is to stabilize neural circuits.

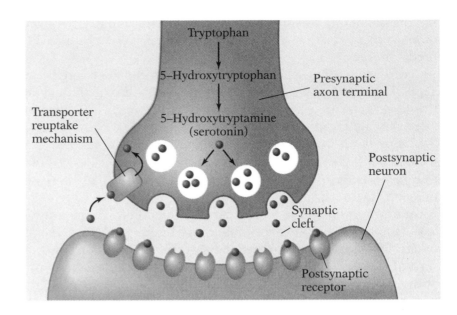

role in these organs as it does in the brain. Serotonin often modulates the response elicited by other neurotransmitters. Serotonin is made from the amino acid tryptophan, which is abundant in meat and fowl. (The human body cannot make tryptophan, and thus we must obtain it from dietary sources. Tryptophan deprivation alters brain chemistry and mood.) Tryptophan is obtained by the digestion of proteins in the gut and is transported in the blood plasma to the brain, where it is converted to serotonin. Serotonin is released from axon terminals and binds to specialized receptors in the membrane of the target neuron. Serotonin is also absorbed from the synaptic cleft by a special transporter reuptake mechanism. As we shall see, the receptors and the reuptake mechanism have important roles in the evolution of the serotonergic system.

If one thinks of the structure of the brain as a house, the serotonergic neurons are located in the basement. Like the basement regulators of water and electricity, this set of neurons is fundamental to the functioning of the house, acting somewhat like the house's thermostat to maintain a comfortable equilibrium in response to outside variations. The cell bodies of the serotonergic neurons occupy virtually the same location in the basement of every vertebrate brain and are even in the same spot in the central nervous system of amphioxus, a primitive chordate. Thus the serotonergic system was essentially in place 500 million years ago, and it has been amazingly conserved throughout evolution, yet it participates vitally in the most complex aspects of our thinking and emotions. The axons of these neurons release serotonin, which generally does not directly excite other neurons but instead *modulates* the responses of neurons to other neurotransmitters. In some instances, however, serotonin does directly excite other neurons, such as the pyramidal neurons in the cerebral cortex.

The axons of the serotonergic neurons project in rich profusion to every part of the central nervous system (the brain and spinal cord), where they influence the activity of virtually every neuron. This widespread influence implies that the serotonergic neurons play a fundamental role in the integration of behavior. Our sense of well-being and our capacity to organize our lives and to relate to others depend profoundly on the functional integrity of the serotonergic system. There are only a few hundred thousand serotonergic neurons in the human brain, roughly *one millionth* of the total population of neurons in the human central nervous system. However, the serotonin receptors on the target neurons are remarkably diverse. Fourteen types of serotonin receptor have been discovered so far in the brains of mammals, located in different places and acting in different ways. These different types of serotonin receptor have a very ancient evolutionary history going back at least 800 million years, and thus some of them came into existence long before brains first appeared about 500 million years ago.

Serotonin receptors are proteins, which are long chains of amino acids that are encoded in the DNA. The sequences of amino acids that make up the receptors have been mapped for more than 30 different members of the family of serotonin receptors present in humans, rats, mice, and fruit flies. The different serotonin receptors were created by a series of gene duplications. (I will have much more to say about the role of gene duplication as an evolutionary mechanism in Chapter 3.) The serotonin receptors are members of a larger family known as the G protein–coupled receptors, which are present even in yeast and molds. Some serotonin receptors are located in the gut or the walls of blood vessels and participate in the regulation of the basic physiological processes of digestion and blood pressure. Most receptor types are located in particular structures within the brain where they appear to regulate the responses of neurons to other neurotransmitters. Overall, the diverse array of serotonin receptors works to achieve a delicate balance in neural activity throughout the central nervous system.

The molecular clock based on serotonin receptors in different organisms. The vertical dimension is a scale of the differences in the amino-acid sequences of serotonin and related receptors between groups of organisms such as mammals versus insects. The horizontal dimension shows the estimated time since the existence of the most recent common ancestor of the members of each of these pairs. The diagonal line expresses the relationships shown by these data.

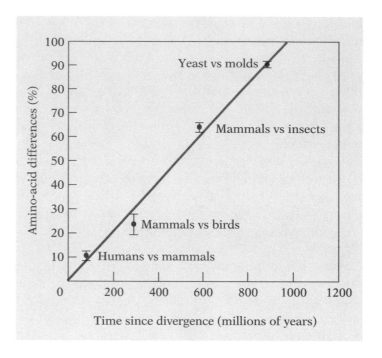

Mutations in the DNA result in changes in the coding for specific amino acids at particular positions in the sequence. These changes seem to occur at a very slow, clocklike rate over millions of years. Thus receptors that diverged in the distant past have a larger percentage of differences in their amino acids than do more recently diverged receptors. The rate of change has been remarkably constant over the past billion years and the times of divergence derived from this measure fit remarkably well with observations based on the fossil record. Thus, for example, the ancestors of humans diverged from nonprimate mammals about 70 million years ago; and the common ancestor of mammals and insects probably existed about 600 million years ago. "Molecular clock" data such as this is extremely useful for establishing the approximate timing of events in brain evolution and has been applied to many other genetic systems.

In an elegant series of experiments with cats, Barry Jacobs showed that the activity of the serotonergic neurons is closely related to the arousal state of the animal. The frequency of the firing of serotonergic neurons declines with decreasing arousal from the active waking state, to quiet waking, to slow wave sleep, and stops entirely in rapid-eye-movement (REM) sleep, when most muscles in the body become inactive. When the animal increases its motor activity, the firing of the serotonergic neurons often increases just *before* the activity begins and continues as long as the motor activity is maintained. Thus the increase in serotonergic neuron activity is apparently driven by the neural commands to move the muscles. The rate of firing of serotonergic neurons often increases with the cat's walking speed on a treadmill. Remarkably similar results have been obtained in recordings from serotonergic neurons in invertebrates such as lobsters and sea slugs, findings that suggest a basic commonality of serotonergic function throughout the animal kingdom. This common function appears to be the stabilization and coordination of neural activity during active movement such as walking, running, or swimming. The relationship between the activation of the serotonergic system and repetitive muscular activity may account for the sense of well-being that many experience following exercise.

Another way to study the serotonergic system is to observe the behavioral changes that result from administering pharmacological

The neural activity of a serotonergic neuron recorded in the brainstem of a cat at different stages of arousal, based on recording experiments by Barry Jacobs. Below the drawings of the cat are records of the neural activity in which individual action potentials are represented by spikes.

Active wake          Quiet wake          Slow-wave sleep          REM sleep

The rate of activity of a serotonergic neuron is related to the cat's movement. When animals engage in a repetitive behavior such as walking on a treadmill, serotonergic neurons increase their activity in proportion to the speed of the activity. The lower cat is walking faster and the serotonergic neuron is firing more rapidly.

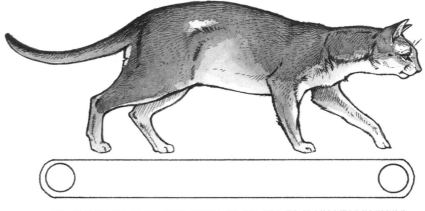

Neural activity

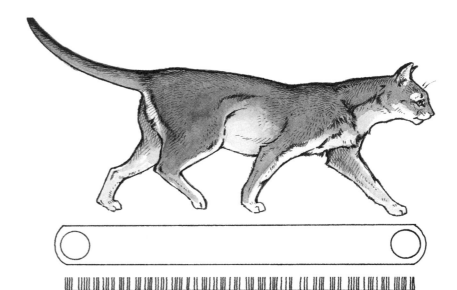

Neural activity

agents that influence the manufacture and reuptake of serotonin. Drugs that *decrease* the amount of serotonin in synapses *increase* exploratory, eating, and sexual behavior, as well as fear-induced aggression. Similarly, when the gene that encodes one class of serotonin receptor is inactivated in mice, the mutant mice are grossly obese and prone to dying from sudden seizures. This evidence also suggests that serotonin constrains the responses of neurons and thus stabilizes the activity of the brain during different behaviors.

These stabilizing constraints result from the influences of the different types of serotonin receptors, each of which has a specific distribution in the brain. The common forms of psychiatric instability such as obsessive-compulsive and anxiety disorders are related to deficiencies in the serotonergic system and are treated by drugs that increase the strength of serotonergic modulation of neural activity.

Michael Raleigh and his colleagues have shown that serotonin is intimately linked to social status in primates. Working with male vervet monkeys in social groups, they found that monkeys with low levels of the serotonin metabolite, 5-HIAA, have low status. Remarkably, they also found that when they manipulated the concentration of serotonin at synapses with drugs, they influenced the monkeys' social standing. (The observers who rated the changes in behavior did not know which drug had been administered.) Thus serotonin levels are not merely a correlate of social status but are directly causal. By contrast, higher status was not related to obvious somatic features such as larger body size or canine teeth. During the course of the experiments, which lasted several weeks, the changes in status were always preceded by changes in affiliative behavior with females. Male monkeys given drugs that increased serotonin engaged in more frequent grooming interactions with females, behavior that was followed by female support in dominance interactions and increased status for the male. Conversely, male monkeys given drugs that decreased serotonin had less frequent grooming interactions with females, and female support in dominance interactions subsequently diminished, resulting in decreased status for the male. The dominant monkeys were more relaxed and confident; the subordinate monkeys were more likely to be irritable and to lash out at other animals.

Raleigh and his colleagues also measured in monkeys the amount of one class of serotonin receptor in the orbital-frontal cortex and the amygdala, brain structures that have an important role in the regulation of social behavior. They found that the amount of this type of serotonin receptor in these structures is strongly positively related to the frequency of prosocial behavior, such as grooming, and negatively related to antisocial behavior, such as fighting. Thus this class of serotonin receptor seems to stabilize the relationships between the individual and other members of its social group.

Why do not all animals have high levels of serotonin and its receptors and live in the most congenial manner possible? The answer may be that low serotonin levels are related to stronger motivational drive and greater sensitivity to rewards and risks in

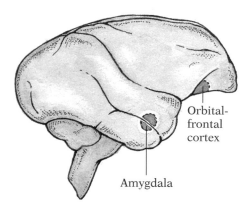

Orbital-
frontal
cortex

Amygdala

Brain structures with important roles in social behavior are the orbital-frontal cortex, located on the lower surface of the frontal lobe, and the amygdala, located deep in the brain in the interior part of the temporal lobe.

the environment. Steven Soumi and Dee Higley have suggested that animals with high serotonin levels, while more stable, are less sensitive to hazards and opportunities in the environment, which may explain why there is a diversity of serotonin levels in natural monkey populations. The low-serotonin monkeys may be the first of their group to find new food sources and may serve as sentinels that detect predators. The evolution of this increased sensitivity to environmental risks and opportunities is analogous to the evolution of specific alarm and food calls that serve to alert other group members, probably close kin sharing many genes in common, to the presence of predators or resources. Such behaviors may endanger an individual but enhance the survival of close relatives and the propagation of genes shared with the individual. The potential adaptive significance of genes for low serotonergic function may explain why mood disorders, which are associated with low serotonin levels and are typically treated by drugs that enhance the concentration of synaptic serotonin, are so prevalent in the human population.

The concentration of serotonin in synapses is also influenced by the action of serotonin transporters, molecules that scavenge serotonin and return it to the presynaptic terminal in the process known as reuptake. The selective serotonin reuptake inhibitors like fluoxetine (Prozac), which increase the concentration of serotonin in synapses by blocking its reabsorption, have become very important drugs for the treatment of depression, anxiety, and obsessive-compulsive disorders. Klaus-Peter Lesch discovered that in monkeys and humans, the serotonin transporter gene is under the control of a special DNA promoter sequence, unique to anthropoid primates, that apparently came into existence about 40 million years ago. Lesch found that variations in the DNA sequence of the transporter promoter are associated with variations in the personality traits of anxiousness, hostility, depression, and impulsiveness in humans.

In summary, then, the function of the serotonergic system is to modulate the strength of connections so as to produce stable neural circuits as the organism engages in a wide variety of different behaviors. This function is so fundamental that the basic architecture of the serotonergic system has been preserved for half a billion years. Reducing the strength of serotonergic modulation increases motivational drive and sensitivity to both risk and reward, which can in some circumstances confer adaptive benefits. However, this increased sensitivity also confers increased vulnerability to a wide variety of dysfunctions that afflict contemporary humans, including anxiety, eating, stress, obsessive-compulsive, and sleep disorders,

## Serotonin, Cholesterol, and Violence

While high levels of serum cholesterol are associated with an increased risk of heart disease, recent epidemiological studies by Beatrice Golumb and others have revealed the disturbing finding that low cholesterol is associated with an increased risk of violent death from accidents and suicide. Experimental studies by Jay Kaplan and his colleagues have found that monkeys fed a low-cholesterol diet are significantly more aggressive and have lower levels of the serotonin metabolite, 5-HIAA, in their cerebrospinal fluid than do monkeys on a high cholesterol diet. Monkeys in both groups received the same amount of calories and were the same body weight. Reduced serotonin leads to increased food-seeking and risk-taking behavior. Cholesterol is required for many functions in the body and is an important constituent of neural membranes. Moreover, cholesterol is typically found in energy-rich animal food sources. From these observations, Kaplan and his colleagues suggest that the linkage between cholesterol and serotonin may have been selectively advantageous in early human populations because it would have enhanced the acquisition and consumption of vital nutrients. For many contemporary human populations, cholesterol- and energy-rich food sources are superabundant; their excessive consumption today is driven by retained adaptations to former conditions in which these food sources were scarce.

substance abuse, and depression. The manifestation of depression might seem in conflict with an underlying mechanism that typically increases motivational drive, but it can be regarded as the exhausted state produced by hypersensitivity. The conditions of contemporary life are far removed from the circumstances in which we evolved. The mechanisms for vigilance that conferred a survival advantage in the evolutionary past may in some cases turn pathological in contemporary life, in which we are flooded with artificial stimuli demanding our attention. Sedentary life-styles and a consequent reduction in the activation of the serotonergic system may also be responsible for increased levels of psychopathology.

# The Neocortex

The cerebral cortex is a sheet of neural tissue covering much of the brain. In contrast to the serotonergic system, which is basically similar in all vertebrates, the neocortex, a part of the cerebral cortex, is a structure found only in mammals. Ranging by a factor of about 100,000 from the tiniest shrews to the whales, neocortex size is related to body mass; however, when the effect of body mass is taken into account relative neocortex size still varies by a factor of more than 125. Other parts of the cerebral cortex do not vary nearly so much as the neocortex. For example, the hippocampus, which has been a favorite subject of investigation by memory researchers, varies by less than a factor of 8 across the same set of mammals in the extensive volumetric studies of Heinz Stephan and his colleagues. Some mammals, such as the primates and the toothed whales, have a much larger neocortex than would be expected for their body mass. The neocortex is a folded sheet of neural tissue a couple of millimeters thick. The unfolded human neocortex would make a fair sized napkin of about 200,000 square millimeters. It is folded into a compact bundle so as to decrease the amount of wiring needed to connect different parts of the sheet and perhaps so that it can fit in a baby's skull small enough to pass through the mother's birth canal.

# Mapping the Neocortex

The word "cortex" means the outer shell or rind of an object. In accordance with its rather prosaic name, the early anatomists did not attach much importance to the structure. In the seventeenth century, the Italian anatomist Marcello Malpighi first examined the cortex with a primitive microscope, and he reported seeing tiny glands that fed into a system of ducts. Malpighi was perhaps inspired by the ancient theory of Hippocrates that the brain secreted phlegm into the nasal cavity. Emanuel Swedenborg, the eighteenth-century Swedish polymath, was the first to appreciate the functional role of the cortex. In 1740, he wrote: "the cortical substance imparts life, that is sensation, perception, understanding and will; and it imparts motion, that is the power of acting in agreement with will and with nature." He believed that the cortical glandules seen by Malpighi were cerebullula ("tiny brains") that were connected to one another

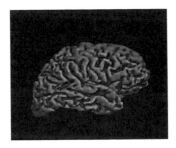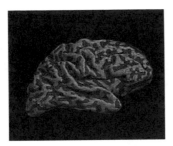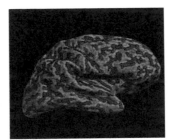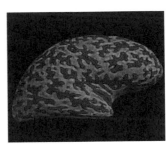

by threadlike fibers, which we now recognize as neurons and axons. He noted that similar tiny fibers arose from the sense organs and terminated in the cortex. Other fibers emerged from the cortex, passed through the underlying white matter, and descended to the spinal cord, where they entered the peripheral nerves and connected the cortex with the muscles of the body. He proposed that the motor functions in the cortex are topographically mapped, with the control of the muscles of the foot located in the dorsal cortex and the control of the muscles of the face in the ventral cortex. In all these conclusions, and in many others that he made about the brain, he was correct, but it would be more than a century before his theories were unwittingly confirmed by other investigators. Sadly, Swedenborg's prescient ideas about cortical functions were largely ignored in his own time and for more than a century afterward.

The opposite was true for the ideas put forth by the early cortical anatomists Franz Josef Gall and Johann Spurzheim, who published their widely read book, *Anatomie et Physiologie du Système Nerveux*, in 1810. They provided the first accurate descriptions of many brain structures, but they are much better known for the idea that the brain is made of specific organs responsible for personality traits such as pride, vanity, humor, benevolence, and tenacity. They believed that they could detect these structures by measuring bumps in the skull, which they thought were produced by the expansion of the underlying brain organs in individuals who strongly exhibited the corresponding traits. Spurzheim coined the term "phrenology" ("mind study") to describe this endeavor. Phrenology was widely rejected by nineteenth-century scientists but embraced by the popular culture. Dozens of societies, presses, and museums devoted to the practice of phrenology sprang up in Europe and

Unfolding the convolutions of the human neocortex. The parts of the neocortex located on the outer surface are shown in green; the buried parts are shown in red. Martin Sereno and Anders Dale created this unfolding on the basis of magnetic resonance images. These and other brain maps can be seen in movie format at http://cogsci.ucsd.edu.

Franz Josef Gall and Johann Spurzheim's 1810 illustration of the human brain, which was one of the earliest to show accurately the convolutions of the neocortex. Earlier depictions of the neocortex look like piles of intestines.

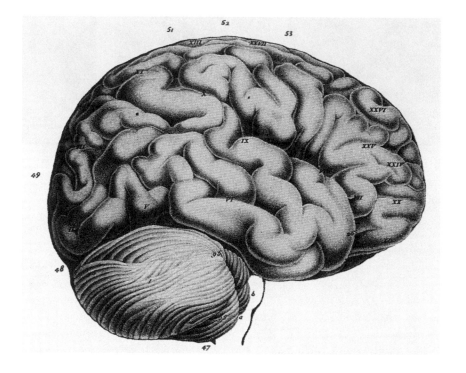

America, and phrenology continues to maintain a hold on the popular imagination even to this day. The phrenological maps are pure fantasy without any basis in experimental or clinical observations. However, the phrenologists can be credited with the general idea that functions are localized in particular places in the brain.

In the early nineteenth century, the French physiologist Pierre Flourens tested the theories of Gall and Spurzheim by removing parts of the brain in animals. Flourens was unable to confirm the phrenological maps, but he did establish the foundations of experimental neurobiology. In 1825, Thomas Jefferson commented on these experiments in a letter to his friend and former political opponent, John Adams: "I have lately been reading the most extraordinary of all books. It is Flourens' experiments on the functions of the nervous system, in vertebrated animals. He takes out the cerebrum compleatly, leaving the cerebellum and other parts of the system uninjured. The animal loses all its senses of hearing, seeing, feeling, smelling, tasting, is totally deprived of will, intelligence, memory, perception, yet lives for months . . . in a state of the most absolute stupidity." Adams replied: "Incision knives will never discover the distinction between matter and spirit. That there is an active principle of power in the Universe is apparent, but in what substance that active principle resides, is past our investigation." Jefferson's enthusiasm reflects the intellectual excitement created by these early efforts to understand brain function, and Adams's response reflects the skepticism and perhaps apprehension evoked by these early studies. The neocortex is the principal component of the cerebrum. Flourens's observations are the first experimental evidence implicating the neocortex in the functions of perception, volition, and memory.

The first definite localization of function within the neocortex was made by the French anatomist Paul Broca in 1861. Broca did a post mortem examination of the brain of a patient named Leborgne, who for 20 years had been able to speak only a single word, "tan." He found a well-defined lesion in the frontal lobe of this patient's brain and concluded that it was responsible for his disability. Broca's localization of the speech area in the frontal lobe has been repeated in many studies of brain-damaged patients, by electrical stimulation of the area during neurosurgery and, more recently, by functional imaging studies showing that this area is active during speech production.

The first part of the neocortex to be topographically mapped was the area involved in the control of the muscles of the body. In the 1860s, the British neurologist John Hughlings Jackson observed that in some epileptic patients a seizure would progress from one part of the body to another. He described what he called the "march of epilepsy" in one patient: "A married woman, 43 years of age, but looking ten years younger, consulted me at the London Hospital, December 13, 1864. Exactly a week before, at 9 or 10 a.m., her right forefinger and thumb began to work [convulse], and the working continued up to the elbow and then all the fingers worked. The fit was strictly localized . . . . She had had three attacks, and after each the hand felt heavy and dead, and for some time she could not use it well." This type of fit, today called a Jacksonian seizure, is also known as a complex partial seizure because it is confined to a particular part of the body. Hughlings Jackson concluded that the muscles were "represented" in the brain in a particular location, which he deduced to be somewhere in the cerebral cortex or in a nearby structure called the corpus striatum. This theory was a radical departure from the prevalent clinical view of the time, which was

An anxious man attempts to use the technique of phrenology to assess his mental capacities; a lithograph, c. 1825, after Theodore Lane.

The prescient clinical observations of
John Hughlings Jackson (1835–1911)
led the way to understanding cortical
organization.

that epileptic seizures were caused by a disturbance in the lowest
level of the brain stem. Hughlings Jackson further noted, "They
rarely begin in the upper arm, or in the calf. The fit usually begins in
that part of the face, of the arm and of the leg which has *the most
varied uses*. Fits beginning in the hand begin usually in the index finger and thumb; fits which begin in the foot begin usually in the great
toe." From this observation he deduced that the parts which have the
most varied uses will be represented in the central nervous system
by the most neurons.

Hughlings Jackson's clinical observations relate to three fundamental properties of the neocortex. The first is that the neocortex
contains topographic maps, the second is that the parts of these
maps which are used the most have the largest representations, and
the third is that the neocortex has a key role in the genesis of
epilepsy. The sites of abnormal tissue that initiate epileptic seizures
are primarily located in the neocortex or in other cortical structures, such as the hippocampus. The cortical circuitry is highly
plastic in that it can change its functional organization in response
to experience, and it is crucial for memory formation and storage.
The price of this cortical plasticity is the risk of the wildly uncon-
trolled oscillations in neural activity that occur in epilepsy. Thus the
risk of epilepsy may be the inevitable cost of the adaptive properties
of the cortical neural circuitry.

In 1870, Hughlings Jackson's topographic prediction was confirmed by the German physicians Gustav Fritsch and Eduard Hitzig,
who discovered the motor cortex by stimulating the surface of the
brain in dogs with weak electrical currents and observing discrete
movements of the body. When they repeated the stimulation at the
same site they observed the same movement; when they stimulated
nearby sites they observed movements in adjacent muscles. The
Scottish neurologist David Ferrier did much more extensive experiments in monkeys and showed that there is a topographically organized map of the muscles of the body in the motor cortex. In 1876,
Ferrier published the first cortical maps in his book, *The Functions
of the Brain*. Subsequently, the neurophysiologists Charles Sherrington and Cecile and Oscar Vogt and the neurosurgeons Otfried Foer-
ster and Wilder Penfield showed that the map in the motor cortex
emphasized the muscles of the hand and face in monkeys, apes,
and humans.

More recently, many investigators have stimulated the motor
cortex with microelectrodes and found fine-grained mosaics in

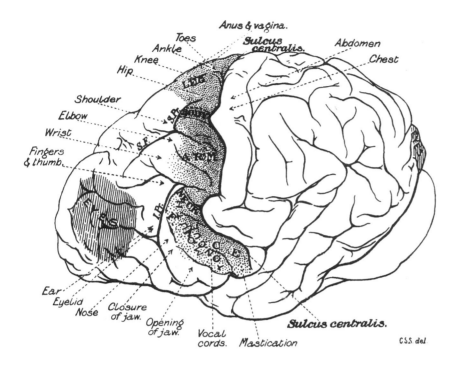

Anus & vagina.
Toes
Ankle
Sulcus
Knee
centralis.
Hip
Abdomen
Chest
Shoulder
Elbow
Wrist
Fingers
& thumb.
Ear
Eyelid
Nose
Closure
of jaw.
Opening
of jaw.
Vocal
cords.
Mastication
Sulcus centralis.
C.S.S. del

The motor cortex in the left hemisphere of the chimpanzee brain, as mapped by A. S. F. Grünbaum and Charles Sherrington in 1902. The hatched area marked "eyes" is the frontal eye field, which receives input from the visual cortical areas such as MT. This area, in conjunction with the optic tectum, controls eye movements.

which different muscles are represented in exquisite detail. Randolph Nudo and his colleagues have found that the size of the motor representation of the fingers depends on experience. They mapped the motor cortex with microelectrodes in a group of squirrel monkeys and then tested the influence of subsequent finger use on the motor maps. One group of monkeys was trained to retrieve food from a small well, a task that required fine control of the individual fingers. A second group of monkeys was trained to retrieve food from a large well, from which they could grasp the food with their whole hand. Nudo and his colleagues then remapped the motor cortex in both subgroups. The monkeys that had performed the fine finger movements had a significantly larger cortical representation of the fingers than they did before training, whereas there was no change in the monkeys whose task could be performed with their whole hand. Functional imaging experiments done in human subjects have also demonstrated that the hand representation expands as a result of performing complex finger movements. The expansion of the hand representation can be observed following short-term

Clinton Nathan Woolsey (1904–1993), photographed at the Laboratory of Neurophysiology of the University of Wisconsin, holding one of his cortical maps.

training, but it is most notable in Braille readers and in musicians who play stringed instruments. These findings demonstrating the role of experience build upon Hughlings Jackson's original observation: the finer the degree of control and use of a muscle, the larger its representation in the cortex.

Thus the muscles that are crucial to manipulating objects, eating food, and making facial expressions have a disproportionately large area of motor cortex devoted to their control relative to the amount of cortex devoted to the weight-bearing muscles used in standing and locomotion, which have a much greater physical mass. Manipulation, mastication, and facial expression require much finer control of individual muscles than does the maintenance of posture and locomotion. There is a dynamic interplay within the motor maps such that the skilled use of particular muscles is associated with the expansion of their cortical representation.

With the development of electronic amplifiers and oscilloscopes in the 1930s it became possible to record the electrical activity of the cortex. Edgar Douglas Adrian, Clinton Woolsey, and their colleagues found that the region adjacent to the motor cortex was electrically activated

Raccoon                                                                Coatimundi

A comparison between maps in the somatosensory cortex of the raccoon and the coatimundi. The representations of the different parts of the body in the brains are shown below. The opposite side of the body is represented in each hemisphere of the brain. The representations of the forepaw, greatly enlarged in the raccoon and much smaller in the coatimundi, are outlined in red. As shown in the enlargement of the forepaw map, Wally Welker and his colleagues also found that the representations of the individual digits of the highly sensitive raccoon forepaw are separated by small fissures; the skin between the digits is less sensitive and is represented in the bottoms of these fissures. Their observation suggests that one mechanism for the formation of cortical fissures results from the differential expansion of the representation of the more sensitive parts of the receptive surface, in this case the skin of the forepaw.

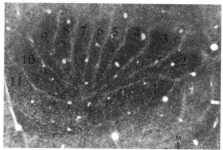

Left: A star-nosed mole. Top right: A close-up view of the appendages on one side of the nose, numbered counterclockwise from 1 to 11. Bottom right: The representation of these appendages in the somatosensory cortex superimposed on a brain section stained for cytochrome oxidase. Note that each appendage corresponds to a separate zone in the mole's somatosensory cortex. The light stripes between the zones contain fewer neural connections than the cortex within each zone. The connections within the representation of each appendage are stronger than those between adjacent appendages. Martin Sereno and I found a similar image of the fingers of the hand in the somatosensory cortex of the owl monkey.

by mechanical stimulation of the surface of the body and named it the somatosensory cortex, from the Greek *soma*, "body." When they recorded from a particular site in the somatosensory cortex, they were able to map out a receptive field on the body surface which activated that site. By systematically moving the recording electrode from point to point on the cortical surface they were able to determine the representation of the body surface in the somatosensory cortex; they also found a second map of the body surface nearby.

In the 1970s, by using microelectrode recordings, Michael Merzenich, Jon Kaas, and their collaborators were able to establish that there are at least four maps of the body surface in the somatosensory cortex of monkeys. Like the motor cortex maps, the somatosensory cortex maps in primates show a strong emphasis on the hand and face, indicating that the exquisitely sensitive surfaces of the hand, lips, and tongue are connected to much larger areas of cortex than are the less sensitive parts of the body. As with the motor cortex, the somatosensory cortical maps are plastic and the cortical representation expands for the parts of the body that are heavily used. The distinction between somatosensory cortex and motor cortex is not absolute. The motor cortex has some sensory functions, and vice versa.

The regions of the body that are behaviorally important for fine movement or discrimination have large cortical representations. This principle was beautifully demonstrated by Wally Welker, who mapped the somatosensory cortex in the raccoon, which relies on the fine sensitivity of its forepaws for the detection of prey, and in its close relative, the coatimundi, which uses a highly sensitive snout rather than its paws to find prey. The raccoon has an enormous representation of the sensitive surface of its forepaws in its somatosensory cortex; by contrast, the coatimundi has a greatly enlarged representation of its sensitive snout. In both cases the enlarged cortical representations correspond to tactile organs used to probe the environments in search of food.

Another beautiful example of the relationship between a sensory specialization and the cortical map is the star-nosed mole. This mole has 22 fingerlike appendages extending from its sensitive nose that it uses to probe its way through its underground system of tunnels. Kenneth Catania and Jon Kaas found a greatly enlarged representation of the separate tactile appendages in the somatosensory cortex of the star-nosed mole.

## Mapping the Visual Cortex

The first steps in the mapping of the visual cortex came about through the tragic circumstances of war. In the Russo-Japanese War of 1905, many Japanese soldiers sustained bullet wounds that penetrated through the posterior part of their brains. Because of the higher muzzle velocity and the smaller bullet size of rifles developed in the late nineteenth century, these weapons tended to produce more localized brain injuries than were inflicted in earlier wars, and improved care of the wounded also resulted in higher rates of survival. Many of the wounded soldiers were partially blinded by these injuries, and Tatsuji Inouye, an ophthalmologist, was asked by the Japanese government to evaluate the extent of their blindness as a means to determine their pension benefits. Inouye found that the parts of the visual field in which these soldiers were blind corresponded to the locations of their brain injuries as determined by the sites of the bullet's entry and exit through the head. By combining the visual field deficits from different soldiers he was able to deduce the topographic organization of the primary visual cortex. Inouye's map revealed that much more cortex was devoted to the representation of the central part of the retina than to the periphery. This is the portion of the

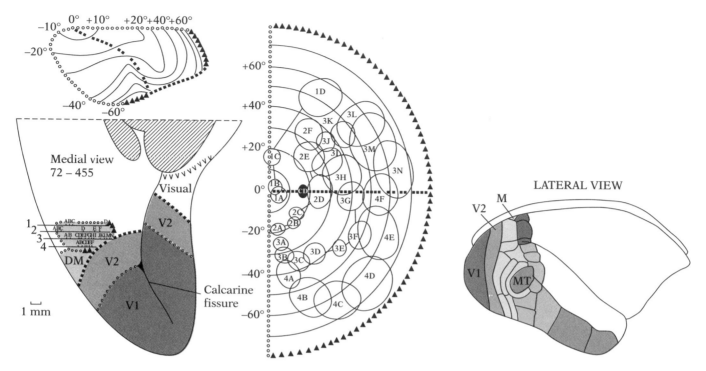

Left: Making a microelectrode map of a small visual area (M; see brain diagram at right) located on the medial wall of the hemisphere in the owl monkey. Four microelectrode penetrations were passed through the cortex on the wall; the recording sites are indicated by letters for each penetration. The sites of the corresponding receptive fields mapped at each recording site are shown in the semicircular chart. For example, in penetration 1, the first receptive field (1A) was located near the center of the visual field, and the receptive fields 1B, 1C, and 1D marched upward in the visual field as the microelectrode was advanced through the area. The receptive fields mapped in the course of this experiment revealed a highly topographic map, illustrated at top left. This map contains an unusually large representation of the more peripheral parts of the visual field. In the other cortical areas the representation of the central visual field is much larger. Right: The cortical visual areas in the owl monkey outlined on the surface of the brain. The primary visual cortex (V1) is red; the second visual area (V2) is orange; the third tier of visual areas is shown in yellow; the middle temporal area (MT) is blue; the inferotemporal areas are green; the temporoparietal areas are lavender; the posterior parietal areas are brown.

retina with the highest acuity, and it is our most important means for probing our environment for information, and the part you are using to read this book. Inouye's map of the primary visual cortex has been confirmed by modern brain-imaging techniques.

In the 1940s, the neurophysiologists Samuel Talbot and Wade Marshall mapped the receptive field organization of the primary visual cortex with large electrodes placed on the cortical surface. Talbot also established that a second visual area was located adjacent to the primary cortex. In the 1960s, Jon Kaas and I began map-

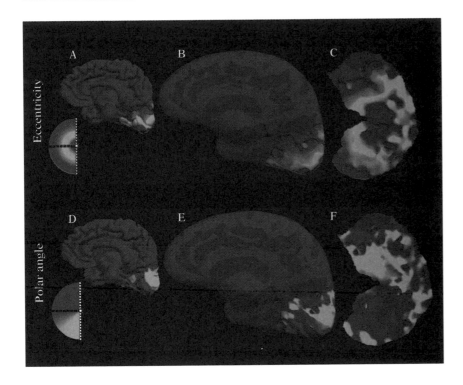

A map of the human visual cortex based on a functional magnetic resonance imaging study by Martin Sereno and his colleagues. Functional MRI is based on measurements of blood oxygen levels, which in turn are linked to local brain activity. Brain activity is evoked by moving stimulus patterns across the visual field of the subject. Images A and D show the medial wall of the right hemisphere of the brain; B and E are the same brains unfolded. In the upper left is a visual field map in which color indicates eccentricity: blue is near the center of the visual field and red is the far periphery of the visual field. The color-coding in images A and B shows the locations of the representations of these parts of the visual field. Image C shows a complete unfolding of the visual cortex with the eccentricity map superimposed. In the lower left is a visual field map in which color indicates polar angle, the angle between the center of the visual field (the fixation point) and a line extending to the periphery of the visual field. The color-coding in images D and E shows the locations of the representations according to polar angle. Image F shows a complete unfolding of the visual cortex with the polar angle map superimposed. (Another map of the human visual cortex is illustrated on page 147.)

ping the visual cortex in monkeys with microelectrodes, which permitted much finer resolution than had the earlier surface electrode mapping techniques. To our amazement we found that the visual cortex was much more extensive than anyone had anticipated, and that there were many cortical visual areas, each with its own map of the visual field. By the 1990s, the number of cortical visual areas discovered in primates had grown to more than two dozen. Martin Sereno, Roger Tootell, and their colleagues have been able to map many of these areas in humans using functional magnetic resonance imaging.

## Why Are There Maps in the Neocortex?

One reason for neocortical maps may be wiring economy. Nontopographic representations of sensory spaces would require longer and denser fiber connections than do topographic representations. Much of the analysis of the visual scene involves comparisons between topographically adjacent features that differ in shape or color, for example. Thus one would expect the richest connectivity to be between topographically adjacent parts of the visual field.

The evolutionary expansion of the size, organization, and number of cortical maps appears to be related to the elaboration of behavioral capacities. For example, opossums and hedgehogs, which in many respects resemble the early mammals that lived more than 60 million years ago, have rather limited visual capacities and a small number of the visual cortical areas. In these mammals, the cortical maps of the retina are relatively uniform in that the amount of cortical space devoted to the more central parts of visual field in front of the animal is not much greater than the cortex devoted to the more peripheral parts of the visual field. By contrast, primates have extremely well developed visual capacities and have a large number of cortical maps devoted to visual perception and memory. Within most of these maps there is a strong emphasis of the representation of the central part of the visual field and a much smaller representation of the peripheral parts of the visual field.

## Building on the Past

In 1971, when we contemplated the emerging evidence that there were many cortical maps, Jon Kaas and I suggested that evolution of cortical areas proceeded by replication of pre-existing areas. We were inspired by the paleontologist William King Gregory, who in 1935 suggested that a major mechanism in evolution has been the replication of body parts due to genetic mutation in a single generation that was then followed in subsequent generations by the gradual divergence of structure and functions of the duplicated parts.

Why are separate cortical areas maintained in evolution? One reason for the retention of older mechanisms occurred to me during a visit to an electrical power-generation plant belonging to a public utility. The plant had been in operation for many decades, and I noticed that there were numerous systems for controlling the generators. There was an array of pneumatic controls, an intricate maze of tiny tubes that opened and closed various valves; there was a system of controls based on vacuum tube technology; and there were several generations of computer-based control systems. All these systems were being used to control processes at the plant. When I asked why the older control systems were still in use, I was told that the demand for the continuous generation of power was too great to allow the plant to be shut down for the complete renovation that would be required to shift to the most up-to-date computer-based control system, and thus there had been a progressive overlay of control technologies, the pneumatic, vacuum tube, and computer systems integrated into one functional system for the generation of electrical power. I realized that the brain has evolved in the same manner as the control systems in this power plant. The brain, like the power plant, can never be shut down and fundamentally reconfigured, even between generations. All the old control systems must remain in place, and new ones with additional capacities are added on and integrated in such a way as to enhance survival. In biological evolution, genetic mutations produce new cortical areas that are like new control systems in the power plant, while the old areas continue to perform their basic functions necessary for the survival of the animal just as the older control systems continue to sustain some of the basic functions of the power plant.

The power plant analogy also applies to the evolution of the multiple serotonin receptors. The multiple receptors with different functions are the result of new genes that were produced by the duplication of pre-existing receptor genes through the span of evolutionary time. Duplication is a fundamental mechanism in the evolution of genes, and I will explore its role in brain development and evolution in the next chapter.

A developing human. Evolution is the product of changes in the genes that regulate development.

# Duplicated Genes and Developing Brains

Each living creature is a complex, not a unit; even
when it appears to be an individual, it nevertheless
remains an aggregation of independent parts,
identical in idea and disposition, but in outward
appearance similar or dissimilar.

Johann Wolfgang von Goethe,
*The Metamorphosis of Plants*, 1790

How do the parts of the brain and the body "know" where they are supposed to be located? How does the embryo "know" when to start forming the various body parts so that the developmental sequence results in a viable organism and not a monster? Some of the most profound discoveries of modern biology have been the master regulatory genes that control the shape and pace of body development in both vertebrates and invertebrates. These master regulators are sets of replicated primordial genes. The processes of replication and diversification of function in the replicated genes are the principal mechanism through which evolutionary change comes about. These great discoveries were anticipated to a remarkable degree by the ideas of some of the philosophers and biologists of the eighteenth and nineteenth centuries.

## Repeating Structures and Archetypes

In the late eighteenth century, the great philosopher and naturalist Johann Wolfgang von Goethe observed that organisms are made of repeating structures sharing the same basic anatomical pattern. These basic structures undergo transformation in different parts of the organism; Goethe proposed, for example, that the petals of flowers are transformed leaves. He also theorized that the skull consists of transformed vertebrae, an idea that occurred to him as he contemplated the skull of a sheep in a Venetian graveyard. His remarkable conjecture is supported by modern studies that have revealed the presence of a reduced skull and extra vertebrae in primitive jawless fish and by genetic experiments in mice that transformed the posterior skull into vertebrae. Goethe was a member of the German school of *Naturphilosophie,* which developed the concept of *Bauplan* ("body plan") to describe the basic features characteristic of particular groups of plants or animals.

At the time of the deposition of the French king, Charles X, in the revolution of 1830, Goethe was visited in Weimar by his friend Frederic Soret. When Soret entered the room, Goethe exclaimed, "What do you think of this great event?" Soret answered that the expulsion of the royal family was only to be expected in the circumstances. Goethe replied: "We do not appear to understand each other, my good friend. I am not speaking of those people at all, but of something entirely different. I am speaking of the contest of the highest importance for science between Cuvier and Geoffroy Saint-Hilaire."

Johann Wolfgang von Goethe (1749–1832) in the Roman Campagna, painted by his friend J. H. W. Tischbein. Goethe formed his theory of plant metamorphosis during this visit to Italy, 1786–1788.

What so excited the author of *Faust* was a series of acrimonious debates at the Académie des Sciences in Paris between the two great comparative anatomists Georges Cuvier and Etienne Geoffroy Saint-Hilaire. Cuvier believed that the major groups of animals, like the vertebrates and the arthropods, were separate creations and fundamentally different. Geoffroy Saint-Hilaire, on the other hand, saw a common plan transcending the vertebrate and invertebrate classification that he based on the anatomical connections among parts of the body, in particular the position and connections of the nervous system and the appendages. In his view, animals exhibited variations on this common plan. He deduced from his observations that the plan for vertebrates is an *inversion* of the plan found in arthropods, the invertebrate phylum comprising crustaceans and insects. This is another remarkable idea from the early nineteenth century that has received much support from modern studies of the genetic regulation of development, but to Cuvier and other particularists of

Richard Owen (1804–1892), caricatured by Frederick Waddy in this drawing called "Riding His Hobby." Owen is depicted on the back of the skeleton of an extinct giant sloth, Megatherium. Owen, one of the foremost comparative anatomists and paleontologists of his time, was superintendent of the natural history collections at the British Museum.

the age, Geoffroy Saint-Hilaire's notion of a transcendent pattern for the structure of the body was anathema.

Inspired by the *Naturphilosophen*, in 1846 the British anatomist Richard Owen proposed an archetypal ancestor for the vertebrates that possessed multiple repeating elements. Owen postulated that the diversity found in fossil and living vertebrates reflected modifications of the basic vertebrate plan made up of these repeating elements, which were the individual vertebrae and associated structures like the ribs.

Owen remarked: "General anatomical science reveals the unity which pervades the diversity, and demonstrates the whole skeleton of man to be the harmonized sum of a series of essentially similar segments, although each segment differs from the other, and all vary from their archetype." To Owen these variations in the archetype emerged from ideas in the mind of God. This view was supplanted by the theory of natural selection, which was developed independently by Charles Darwin and Alfred Russel Wallace in the late 1850s. The genesis of the theory of natural selection was heavily influenced by the capitalist economic thinking of the early nineteenth century, the essence of which was expressed by the eminent Victorian philosopher Herbert Spencer, who coined the phrase "the survival of the fittest." Contrary to the popular image of Darwin as a retiring intellectual, he was a very successful capitalist. The Darwin–Wallace theory proposed that new species of animals arose from naturally variant forms that were able to produce more offspring, thus allowing them to replace animals with less successful variations that were competing for the same space and resources. Both Darwin and Wallace were also inspired by the enormous diversity in animal life that they observed during their extensive travels in the tropics. Darwin was further influenced by his contacts with plant and animal breeders, who selectively bred domesticated forms that possessed desirable traits in order to improve their stocks. The breeder's artificial selection for desired traits was analogous to the natural selection of traits leading to the increased survival of offspring.

## Variations, Transformations, and Evolution

One of the major difficulties with the theory of natural selection was the problem of explaining major changes in anatomical structure during the course of evolution. The theory explained how small changes occurred, but how did radical changes such as the emergence of new types of animals come about? In his *Materials for the Study of Variation Treated with Especial Regard to Discontinuity in the Origin of Species* (1894), William Bateson amassed evidence from throughout the animal kingdom for the transformation of repeating structures that he believed could be a basis for the emergence of new species. Bateson created the term "homeosis," which he derived from the Greek *homoios*, "like," to express the process of making two things similar. (He had a gift for developing concepts and creating useful new terminology; he also coined the name "genetics" for the scientific study of heredity.) Bateson proposed a means for large evolutionary changes to occur: "The discontinuity of species results from the discontinuity of variation. Discontinuity results from the fact that bodies of living things are made of repeated parts . . . variation in numbers of parts is often integral, and thus discontinuous . . . . [A structure] may suddenly appear in the likeness of some other member of the series, assuming at one

Charles Darwin (1809–1882), co-conceiver with Alfred Russel Wallace of the theory of natural selection. Darwin was 40 at the time of this picture.

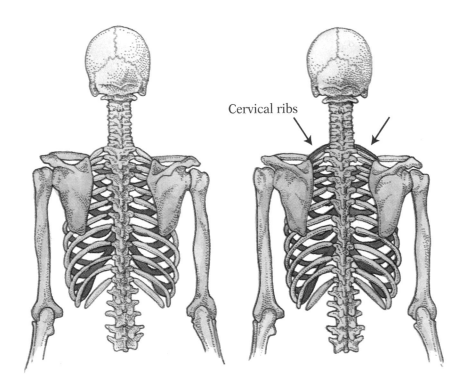

Cervical ribs

A homeotic transformation in humans. On the left is a normal skeleton; in the skeleton on the right are ribs, shown in orange, attached to the lowest cervical vertebra.

step the condition to which the member copied attained presumably by a long course of evolution."

Prime examples of repeating structures are the vertebrae. Bateson noted that the different types of vertebrae underwent homeotic transformation. For example, normally the thoracic vertebrae are connected to ribs, but the cervical vertebrae of the neck do not normally have ribs. However, in the skeletons of rare humans, the lower cervical vertebrae do have ribs or in other cases the upper thoracic vertebrae fail to have ribs. Thus he found evidence that the transformation of vertebrae and ribs could occur in either direction: thoracic into cervical or cervical into thoracic. The concept of evolution through the transformation of replicated elements is applicable both to duplicated genes and to duplicated cortical areas. Key discoveries inspired by Bateson's concept of homeosis were made by geneticists who found the genes responsible for homeotic transformations.

## Gene Duplications and Homeotic Genes

The first physical mapping of a gene to a specific location in the chromosomes was accomplished by the American geneticist Thomas Hunt Morgan and his colleagues in the early 1900s. Morgan used ordinary fruit flies, which provided two enormous advantages for gene mapping: fruit flies reproduce very rapidly, and their salivary glands possess giant chromosomes, 150 times the size of chromosomes in other cells. The microscopic examination of the salivary gland chromosomes revealed a detailed pattern of transverse bands of different thicknesses and structures that could be related to the presence or absence of specific mutations at particular locations.

In 1915, Calvin Bridges, Morgan's colleague in the famous "fly room" at Columbia University, discovered the first homeotic gene mutation in the fruit fly. This mutation transformed the third thoracic segment into the second segment in a manner analogous to the homeotic transformations of the vertebrae described by Bateson. In a normal (wild-type) fly, the second thoracic segment has wings and the third thoracic segment has a pair of sensory structures called halteres. In Bridges's mutation, the halteres were transformed into wings and thus he produced a fly with four wings. Eventually many other homeotic mutations were discovered affecting other parts of the body in fruit flies.

Thomas Hunt Morgan (1866–1945) in the "fly room" at Columbia University in 1918. Morgan founded the Division of Biology at the California Institute of Technology in 1928 and was awarded a Nobel prize in 1933 for his work on chromosomes and heredity.

In 1918, Bridges reported repeated sets of banding patterns in the chromosome map, which presumably contained repeated sets of genes. He proposed that the duplications offered "a method for evolutionary increase in the lengths of chromosomes with identical genes which could subsequently mutate separately and diversify their effects." The geneticist Edward B. Lewis of the California Institute of Technology discovered and mapped additional homeotic genes and conceived the idea that they were duplicates of a primordial gene regulating the development of the body. He also extended Bridges's idea by pointing out that the duplicated gene escapes the

Calvin Bridges (1889–1938), who discovered the first homeotic gene mutation in the fruit fly, in the "fly room" in 1918.

A homeotic transformation in fruit flies. The fly shown at the top is the normal, or "wild-type," fly. The bottom fly has a genetic mutation that has resulted in an extra set of wings.

pressures of natural selection operating on the original gene and thereby can accumulate mutations that enable the new gene to perform previously nonexistent functions, while the old gene continues to perform its original and presumably vital role. As Lewis put it, "A gene which mutates to a new function should, in general, lose its ability to produce its former product, or suffer an impairment in that ability. Since it is unlikely that this old function will usually be an entirely dispensable one from the standpoint of the evolutionary survival of the organism, it follows that the new gene will tend to be lost before it can be tried out, unless, as a result of establishment of a duplication, the old gene has been retained to carry out the old function. The establishment of chromosomal duplications would offer a reservoir of extra genes from which new ones might arise."

An existing gene can sustain only mutations that leave its basic functions intact, otherwise the organism dies without leaving any descendants. However, if that original gene is copied, then the copy can undergo profound mutations while the original gene continues to perform its essential functions. The mutated copy of the original gene can then be influenced by the processes of natural selection in subsequent generations to assume new functions. Thus it becomes possible to link the mechanism of gene duplication to the concept of homeotic genes. In very ancient animals there must have been a

Edward B. Lewis surrounded by mutant flies. Lewis shared in a Nobel prize in 1995 for work on the genetic control of embryonic development.

series of duplications of a primordial gene controlling the development of the whole body. Eventually each member of the replicated array became specialized for the control of the development of a particular part of the body.

Lewis discovered that the genes controlling the development of the thoracic and abdominal segments were located in the same order in the chromosome as the topographic order of the body parts whose development they controlled. Other investigators discovered sets of

The sites of expression of the homeotic series of genes in the fruit fly embryo (above) and the mouse embryo (below). The nose (anterior) is to the left and the tail (posterior) to the right. The genes are expressed in the embryos in the same nose-to-tail arrangement as their order in the chromosomes. The homeotic series are replicated in four linear sets in the mouse chromosomes. The illustrated set (*Hox-2*) is from chromosome 11. Similar linear patterns have been obtained for the sets located on the other chromosomes in mice. In the fly the single homeotic series is broken between genes 6 and 7 into two linear subsets. The common ancestor of mice and flies had a single continuous set. The anterior border of the expression of each gene matches the corresponding color for that gene. The posterior border for each gene is not shown because it overlaps with the anterior border of the next most posterior gene. The evolution of those genes is illustrated in Chapter 4.

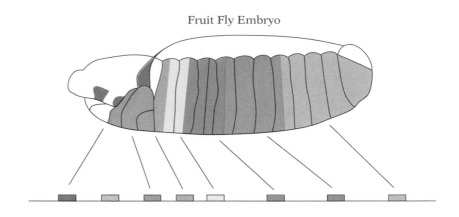

Fruit Fly Embryo

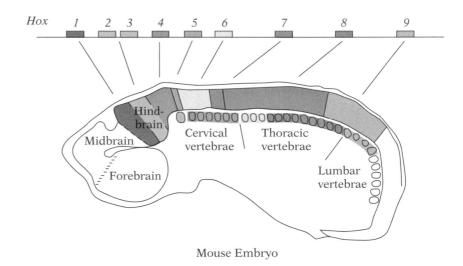

Mouse Embryo

homeotic genes controlling head development in the fruit fly. The discovery of the chemical basis of the genetic code by Marshall Nirenberg and Gobind Khorana led to the development in the 1970s of the techniques for mapping the DNA sequences of the homeotic genes. Lewis's proposal that the homeotic genes in fruit flies were modified replicas led Walter Gehring and his colleagues to search for common DNA sequences among the different homeotic genes. In

1984, William McGinnis and his colleagues and M. P. Scott and A. J. Weiner independently discovered the homeobox, a 180-base-pair DNA sequence that was common to the homeotic genes in the fruit fly. The homeobox encodes a sequence of 60 amino acids, called the homeodomain, that forms the part of proteins that binds to specific sequences of DNA. This general class of DNA binding proteins are called transcription factors. These specific binding mechanisms regulate where and when genes are "turned on" in the developing embryo. The homeodomain is like a hand that slides along the DNA searching for specific sequences and turning on or off genes within a particular part of the body. A gene which is turned on in a particular part of the body is said to be expressed there. Homeobox sequences were soon found in genes throughout the animal kingdom: in hydra, planaria, sea urchins, nematodes, beetles, locusts, amphioxus, fish, frogs, chickens, mice, and humans. These genes were all products of gene duplications at different times in the evolutionary past and were derived ultimately from a primordial gene in the common ancestor of all these animals that contained the homeobox DNA sequence.

## Making the Nervous System from the Neural Tube

The central nervous system forms from a long tube in vertebrate embryos. The major components of the brain—the forebrain, the midbrain, and the hindbrain—are bulges in the anterior parts of the neural tube. The posterior part of the tube becomes the spinal cord. Brain evolution arises from the differential enlargement of parts of the neural tube. Some parts of the tube have expanded greatly in particular groups of animals, and some of the genes that may be responsible have been identified. For example, the roof of the most anterior bulge, the forebrain, becomes the neocortex, which enlarges enormously in primates. One of the genes that may be responsible for the differential enlargement of the forebrain, known as *BF-1*, will be described later in this chapter.

Part of the roof of the hindbrain becomes the cerebellum, which in mormyrid fish expands tremendously to cover the entire brain. In other vertebrates, the formation of the cerebellum is under the control of a pair of genes, *En-1* and *En-2*, that are closely related to a

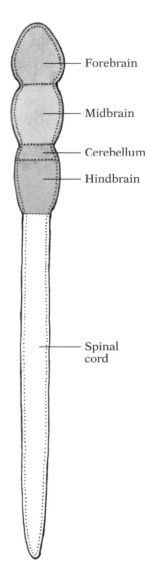

Forebrain

Midbrain

Cerebellum

Hindbrain

Spinal cord

The components of the neural tube that form the precursors of the central nervous system in vertebrates.

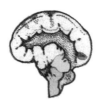
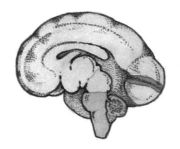
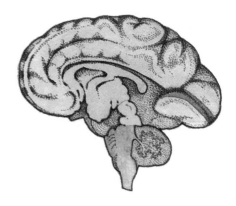

10 Weeks                14 Weeks                    22 Weeks                                28 Weeks

The development of the human brain from 10 to 40 weeks after conception. These are midline views of bisected brains. The calcarine fissure in the primary visual cortex is highlighted in red. The color-coding of the parts of the brain is the same as in the figures on pages 53 and 56.

gene in fruit flies, known as *engrailed,* which contains the homeobox sequence but is not part of the set of homeotic genes that are arranged in somatotopic order in the chromosomes. The great enlargement of the cerebellum in mormyrids may be related to the actions of genes of the *engrailed* family. Thus the differential enlargements of parts of the neural tube in embryogenesis are important factors in brain evolution, and these differential effects are probably the consequence of the actions of homeotic genes.

## Making the Segments of the Hindbrain

The role of the homeotic genes in the formation of brain structures and their connections has been best established in the hindbrain. The vertebrate hindbrain is organized in a series of repeating elements, the rhombomeres. (The name was taken from the rhomboid shape of the hindbrain and *meroi,* the Greek word for "parts.") The rhombomeres were first described in 1828, by the embryologist Karl Ernst von Baer, as a series of ridges across the developing hindbrain. Although the sub-

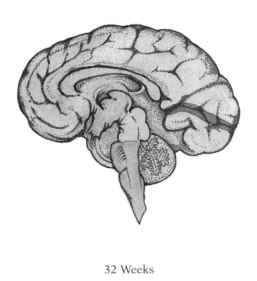

32 Weeks

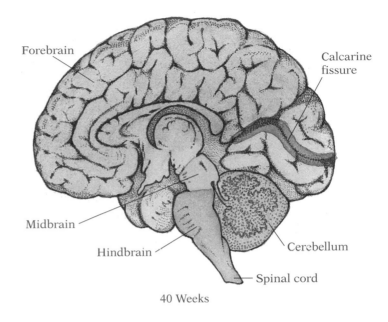

Forebrain

Calcarine fissure

Midbrain

Hindbrain

Cerebellum

Spinal cord

40 Weeks

ject of much controversy during the next century and a half, the rhombomeres are now confirmed as distinct repeating elements. Their segmental identity is linked to the expression patterns of the homeobox genes in the linear order that they occur in the chromosomes. The anterior limits of expression of homeobox genes determine borders between adjacent rhombomeres. Scott Fraser and his colleagues showed that the borders of each rhombomere become barriers to cell migration during development, and thus they form separate compartments. Each rhombomere has a distinct structure and pattern of connections in the adult animal. For example, the fourth rhombomere contains the root of the eighth cranial nerve, which connects the sense organs for hearing and balance with the brain.

The expression of the homeobox genes in the hindbrain and spinal cord is controlled by the concentration of a chemical signal, retinoic acid. Retinoic acid is the biologically active form of vitamin A and is a morphogen, a chemical substance that diffuses through the embryo and controls the spatial and temporal ordering of development. Retinoic acid acts like a hormone, and it binds to specialized receptors that belong to the same family as the receptors for thyroid hormone, which also regulate gene expression. The most

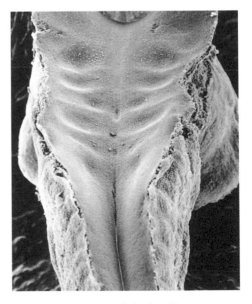

The rhombomeres of the hindbrain viewed with a scanning electron microscope. The overlying cerebellum has been removed.

The expression patterns of some of the major regulatory genes in the developing vertebrate brain. The brain is illustrated schematically as a long tube.

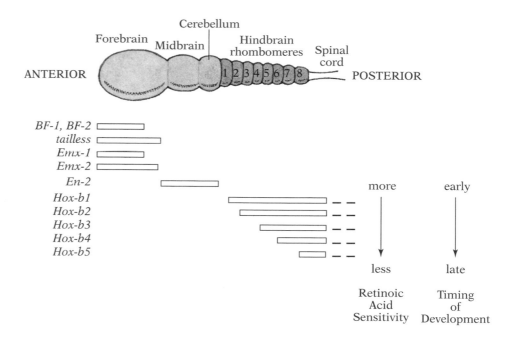

anteriorly expressed homeobox gene in the hindbrain is the most sensitive to retinoic acid, and the sensitivity decreases in a stepwise fashion toward the spinal cord. Thus the genes are turned on from anterior to posterior in a stepwise fashion as the concentration of retinoic acid is increased. The experimental application of retinoic acid in the developing embryo has the effect of advancing the expression of particular homeobox genes and transforming anterior rhombomeres into the pattern normally found more posteriorly. Retinoic acid has a similar role in the developing eye as well as in the limbs and other structures. Because retinoic acid is a morphogen, the ingestion of excessive amounts of vitamin A by pregnant women can alter the development of the fetus and produce severe birth defects.

## Making a Head

The somatopically ordered set of homeobox genes is expressed in linear order in the spinal cord and hindbrain up to the third rhombomere and controls the development of these structures. The more anterior parts of the head and brain are under the control of different sets of master regulatory genes, many of which also contain homeobox DNA sequences but are not part of the somatopic set of homeobox genes. These other sets are also the product of replications of ancient homeobox genes, and it has been possible to establish the lineage relationship of these different sets. William Shawlot and Richard Behringer found that the formation of the entire head anterior to the third rhombomere is dependent on the action of the homeobox gene *Lim-1*. When *Lim-1* is deleted, the entire head anterior to the third rhombomere fails to develop, but the rest of the body is intact.

One of the more remarkable recent discoveries is that the genes that control head and brain formation in fruit flies are very closely related to the genes that control the formation of the more anterior parts of the brain in mammals. One of the most striking of these is the gene *empty spiracles*, which controls the formation of part of the brain in fruit flies. In mammals this gene is replicated, and the duplicas are known as *Emx-1* and *Emx-2*. These genes regulate the formation of the cerebral cortex, and thus the most progressive part of the mammalian brain is controlled by genes with very ancient antecedents going back at least half a billion years. John Rubenstein and his colleagues found a mutation of *Emx-1* in mice that disrupts the formation of the corpus callosum, the major fiber pathway connecting the right and left halves of the cerebral cortex. The corpus callosum is a phylogenetically recent structure present only in placental mammals, clearly demonstrating that old genes can serve new functions in brain evolution. Edoardo Boncinelli and his colleagues found a mutation of *Emx-2* in humans that causes the formation of deep clefts in the cerebral cortex, a condition called schizencephaly, or literally a split brain.

## Brains and Guts

In most animals the brain is located near the entrance to the gut. In vertebrates it is located just above the mouth. In arthropods and molluscs, the brain actually surrounds the esophagus. The consistent location of the brain near the entrance to the gut suggests that the brain arose as the gut's way of controlling its intake by accepting nutritious foods and rejecting toxins. There are several families of genes that govern both brain and gut development, and these

The effect of deleting *BF-1* on the development of mouse embryos. The forebrain is stained deep blue. The embryo on the right is a normal mouse; in the embryo on the left, *BF-1* has been deleted and the forebrain is greatly reduced in size. The replication of neural progenitor cells is greatly reduced in mice lacking *BF-1*.

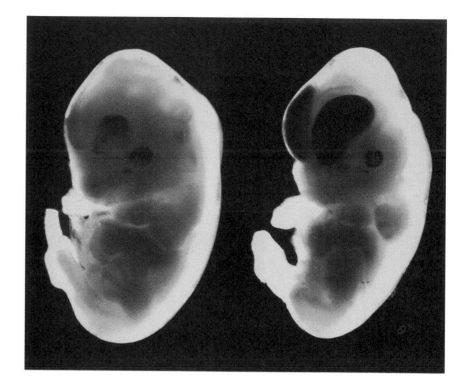

genes may reflect the ancient relationship between gut and brain. In fruit flies, there are two genes, *fork head* and *tailless*, that control development in both the anterior and posterior ends of the embryo and thus are called terminal genes. *Fork head* controls the development of both ends of the gut in flies. Gerd Jurgens and Detlev Weigel found that the *fork head* mutation causes transformations of both ends of the gut into bizarre head structures. Eseng Lai, Wufan Tao, and their colleagues discovered that *fork head* is homologous with the members of a family of replicated genes in mice that is typically expressed in tissues derived from the gut endoderm, such as the liver, lung, and intestines. However, these investigators found that two members of this family, *BF-1* and *BF-2*, are expressed exclusively in the developing forebrain. *BF-1* is expressed in the dorsal half of the forebrain and the nasal retina (the side closer to the nose); *BF-2* is expressed in the ventral half of the forebrain and the temporal retina (the side away from the nose).

Thus the mammalian homologues of *fork head* are examples of replicated genes that have specialized domains of expression and participate in the control of the development of novel structures, in this case the mammalian forebrain. *BF-1* regulates cell division in the germinal zone, which contains the dividing cells before they migrate to form the cerebral cortex, and this gene may control the expansion of the cerebral cortex in primates and other groups of mammals in which the cortex is enlarged. David Kornack and Pasko Rakic have recently shown that the production of cells that form the cerebral cortex in monkeys requires 28 successive rounds of cell division in the germinal zone, whereas the much smaller cortex in mice is produced by only 11 cycles of cell division.

Another example of a gene that controls both ends of the embryo in fruit flies is the gene called *tailless*, but recently Paula Monaghan and her colleagues found *tailless* expressed only in the anterior nervous system in mice. This gene is expressed in the developing forebrain, retina, and olfactory epithelium and thus has a similar distribution to that of *BF-1* and may regulate this gene. A mutation that inactivates *tailless* results in a reduction in the size of parts of the cerebral cortex and amygdala. Thus the terminal genes, which had the ancient function of controlling the formation of the gut, are responsible for controlling the growth of the forebrain in mammals. I will revisit the interesting relationship between the brain and gut in Chapter 7, because there is strong evidence that brain and gut compete for metabolic energy in the organism and that gut size limits brain size.

## Bad Copies

Gene duplications provide the raw material for evolutionary change. However, many mutations of duplicate genes have negative consequences for the organism, such as the mutation of *Emx-1*, which results in the failure of the corpus callosum to develop, and the mutation of *Emx-2*, which produces split brains. Another example with major clinical significance in humans is the mutation that produces spinal muscular atrophy, one of the most common genetic diseases in children. This mutation causes the death of spinal motor neurons and the progressive weakness of the muscles. The mutation occurs in one of two nearly identical *survival motor neuron* genes. Christine DiDonato and her colleagues found only one copy of the *survival motor neuron* gene in mice, which together with the nearly identical structure of the two genes in humans suggests that the duplication occurred relatively recently in evolutionary time. Julie Korenberg has recently proposed that many human diseases arise from mutated gene duplicas; this is yet another instance in which the price of the capacity for evolutionary change is increased vulnerability to disease.

## The Regulation of Development in Space and Time

The spatial ordering of the main set of homeotic genes has been conserved for more than half a billion years. Denis Duboule has proposed that the colinearity in gene location in the chromosomes and gene expression in the body are related to the timing of development in vertebrates. Increasing concentrations of retinoic acid recruit successively more posterior homeobox genes, which generates the timing gradient in which more posterior parts of the body (those toward the tail) develop later than anterior parts. Stephen Jay Gould has proposed that major changes in evolution come about through changes in the timing of development. Changes in the developmental timetable were crucial factors in the origin of mammals as a group, in the origin of primates, and later in the origin of humans. The expression patterns of the homeobox genes indicate that structures in the central nervous system and other parts of the body are regulated by the same genes. For example, some of the same homeobox genes that control brain development also control the development of other cranial structures, such as the jaws and teeth. The tandem evolution of forebrain, jaws, and teeth is the hallmark of the major evolutionary transformations that occurred in the earliest mammals, in the early primates, and again in early humans.

Both replicated structures and replicated genes have the capacity to undergo changes over the generations that enable them to perform new functions while the original structure or gene continues to perform its basic function necessary for the survival of the organism. Thus the duplications provide the raw material for evolution. The lineage history of the replication of genes to form families and subfamilies can be traced by comparing their DNA sequences. In this way evolutionary history can be reconstructed from the DNA of living organisms. However, this approach to understanding brain evolution is an enormous undertaking. The genes described in this chapter are a small part of the vast multilayered network of regulatory genes controlling the development of the nervous system. Nevertheless, from what has been achieved thus far, there is cause for optimism. Two remarkable examples of the role of gene duplication in brain evolution are the homeobox series that controls the specification of repeating structures in the hindbrain, and the duplicate genes *BF-1* and *BF-2*, which control the growth and proliferation of neurons that form the dorsal and ventral forebrain. These later genes, which regulate the most phylogenetically progressive parts of the brain, have exceedingly humble origins among genes controlling the formation of the gut in ancient organisms.

## Ancient Assassins

Brain structures are also shaped by genes that destroy neurons and their progenitors. A clear example of the role of programmed death is the formation of the ganglion cell gradient in the human retina. The retinal ganglion cells send their axons from the retina to the brain. In the adult there is a strong ascending gradient in ganglion cell density from the periphery to the center, and this increasing concentration is partly responsible for the enlarged representation of the central retina in the brain. In the early developmental stages the density of retinal ganglion cells is uniformly high throughout the retina, but as development proceeds many of the peripheral ganglion cells die, creating the gradient. The mechanisms of programmed cell death have been discovered recently. Cell death results from the awakening of dormant "assassins," which proceed to cleave proteins and wreak havoc within the cell. The assassins are members of an ancient family of enzymes known as the caspases, which are the product of a series of gene duplications. In addition to their role in shaping the developing brain and other organs, the caspases have recently been implicated in the destruction of cortical neurons in Alzheimer's disease. Recently Enrica Migliaccio and her colleagues have discovered in gene knockout experiments that removing a trigger for these assassins can extend the lifespan of mice by 30 percent. On page 103, I will explore the possibility that senescence is adaptive and might be an additional stage in the development of an animal.

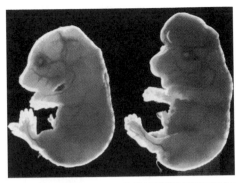

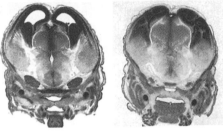

Pasko Rakic, Richard Flavell, and their colleagues have found that mice lacking the cell-death gene, *Caspase-9*, have enlarged brains. The left photographs show normal mouse embryos. In the top right photograph of a mouse embryo lacking *Caspase-9*, the brain protrudes through the skull. In these mice the cortical thickness does not increase but the cortical surface area expands by 25 percent, resulting in the formation of the cortical folds and fissures seen in the lower right section through the brain of a mouse lacking *Caspase-9*. The effect of deleting *Caspase-9* appears to be specific to the brain since the mice exhibited no obvious abnormalities in other parts of the body. Enlarged brains do not help these mice, and they die shortly after birth. A well-functioning brain results from the appropriate expression of genes for cell proliferation and destruction.

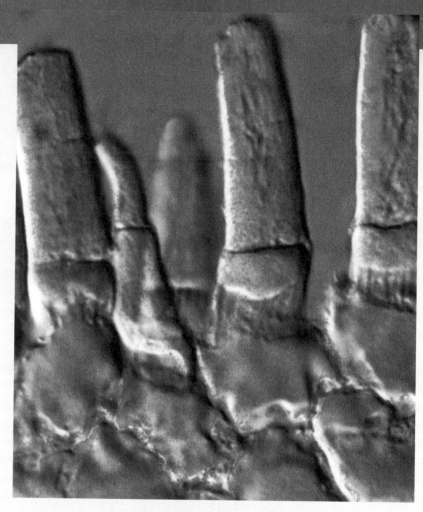

Vertebrate photoreceptors—three rods and a cone in the retina of a salamander—imaged with a confocal microscope. Rods are sensitive to dim light; cones, which require more light, are the basis for color vision.

# Eyes, Noses, and Brains

To suppose that the eye, with all its inimitable contrivances for adjusting the focus to different distances, for admitting different amounts of light, and for the correction of spherical and chromatic aberration, could have been formed by natural selection, seems, I freely confess, absurd in the highest possible degree. Yet reason tells me, that if numerous gradations from a perfect and complex eye to one very imperfect and simple, each grade being useful to its possessor, can be shown to exist; and if further, the eye does vary ever so slightly; and if any of the variations be inherited; and if any of the variations be ever useful to an animal under changing conditions of life, then the difficulty of believing that a perfect and complex eye could be formed by natural selection, though insuperable by our imagination, can hardly be considered real.

Charles Darwin,
*The Origin of Species,* 1859

The evolution of the axon and the action potential enabled neurons to communicate over distances of many centimeters, which in turn made possible the evolution of large and complex animals. These animals first appeared during a period of the earth's history that has been called the Cambrian explosion due to the sudden abundance

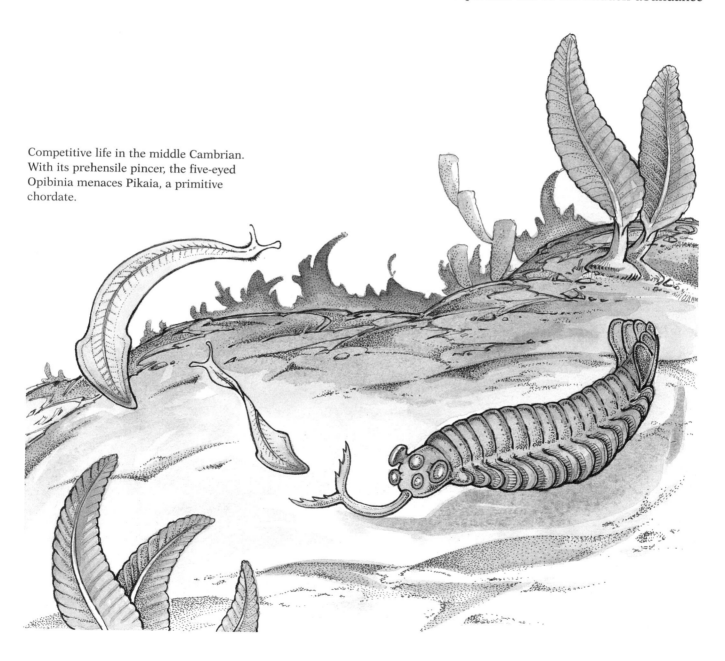

Competitive life in the middle Cambrian. With its prehensile pincer, the five-eyed Opibinia menaces Pikaia, a primitive chordate.

and diversity of fossils. Mark and Dianna McMenamin have proposed that many of these early animals were predatory and the emergence of brains was part of a evolutionary arms race in which different animals struggled for selective advantage. In addition to developing more sophisticated nervous systems, many of these animals acquired body armor as well, and thus they supplanted their brainless, soft-bodied predecessors.

Since Cuvier, it has been recognized that the major groups of animals, the phyla, are differentiated by variations in the basic structure of the nervous system. For example, the bilaterally symmetric pattern of the chordates, like fish, contrasts with the radially symmetric pattern of echinoderms, like starfish. This diversification of phyla can be regarded as natural experiments with different patterns of neural circuitry. The predecessors of most of the living phyla of the animal kingdom arose during the early Cambrian. There were no land animals yet, but in the Cambrian waters swam the first arthropods, ancestors of today's insects, spiders, and crabs; the first molluscs, ancestors of the snails, clams, and squid; the first annelid worms, ancestors of today's earthworms; and our own ancestors, the chordates. In this great proliferation of Cambrian life were also strange animals that have left no descendants living today.

One of the greatest of all biological mysteries is why the diversity of living things suddenly exploded at this time. One hypothesis links a series of massive oscillations in climate with the period of maximum diversification that occurred during the interval between about 540 million and 520 million years ago. By analyzing the magnetic orientation of Cambrian rocks, Joseph Kirschvink and his colleagues have shown that during this short interval there was a 90-degree shift in the orientation of the poles of the earth relative to the continental landmasses. This massive polar shift resulted in global oscillations in climate, which provided new niches for evolving species. The marine sediments from this interval reveal a series of 10 enormous oscillations in the amount of carbon 13, an indicator of the aggregate amount of plant and animal life. These huge Cambrian oscillations exceeded the magnitude of the mass extinction event at the end of the Cretaceous period 65 million years ago, which resulted in the destruction of 75 percent of the animal species then living and led to the emergence of mammals in the following period. During the Cambrian explosion, 10 rapid proliferations of life were followed in each case by a rapid reduction as habitats were created and destroyed. The fossil record shows that new types

A primitive cephalopod, similar to an uncoiled version of the modern nautilus, chases an arthropod in the late Cambrian.

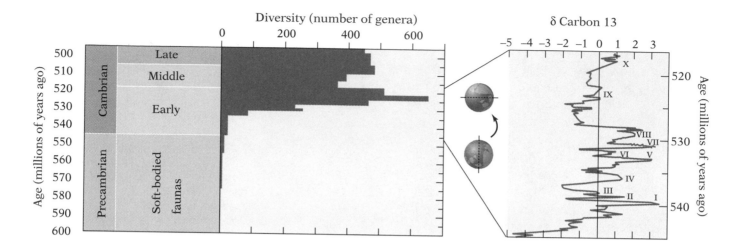

The Cambrian explosion. The graph on the left illustrates the enormous increase in the global diversity of life-forms during the early Cambrian period. The graph on the right is a plot on an expanded time scale showing the oscillations in the amount of carbon 13 as recorded from sedimentary deposits in Siberia. The carbon 13 values are relative to a standard; positive values indicate an increase in biomass. The 90-degree shift in the poles during this period represents a change relative to the continental landmasses, not to the solar system.

of animals appeared during these phases, and so these massive oscillations may have driven the rapid evolutionary diversification that occurred during the Cambrian. For example, the first arthropod fossils appeared in phase V and the first echinoderms in phase VI. There is evidence, however, that the precursors of the different animal phyla existed before the Cambrian. This evidence is based on microfossils of very small organisms that have been recovered in an extraordinary state of preservation from a site in China which is 570 million years old and from estimates of the divergence time of the phyla based on comparisons in the DNA of living organisms. Following the Cambrian, highly developed sensory systems and large brains have evolved independently in two major groups: the vertebrates, within the chordate phylum, and the cephalopods, within the mollusc phylum.

## The Early Evolution of Eyes

The Cambrian animals lived in water, and a fundamental constraint on the evolution of sensory systems in animals living in water is the relative opacity of water to signals throughout most of the electromagnetic spectrum. Eyes sense a small part of the total spectrum, which corresponds to a narrow transparent window we know as visible light. The other part of the spectrum that is relatively transparent in water is in the low frequencies, which are

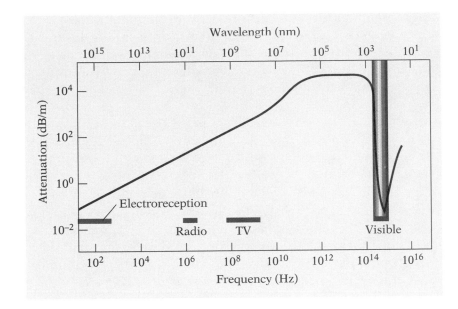

The attenuation (meas[...]
per meter) of electrom[...]
in seawater as a functi[...]
(measured in hertz, cy[...]
and wavelength (meas[...]
Russell Fernald has pointed out that this
physical limitation constrained the early
evolution of photoreceptors in vertebrates
because they lived in water. The later
evolution of vision in vertebrates appears
to have been also constrained by this early
adaptation, because photoreceptors in
vertebrates living outside water have
generally been limited to this range of
the electromagnetic spectrum as well.

used by the mormyrid and other electric fish to probe and sense their environments.

More than 500 million years ago the gene for the photoreceptor protein duplicated and the copies diverged in function. One gene produced a photoreceptor that was sensitive to low levels of illumination, the antecedent of the rod-type photoreceptors that allow us to see in dim light. The second gene produced a photoreceptor that required brighter illumination, the antecedent of our cone-type photoreceptor. During the course of evolution the gene for the cone-type photoreceptor has undergone further duplications that produced proteins that varied in their sensitivity to different parts of the visible spectrum. This is the basis for color vision, and it has evolved separately in different groups of animals. Part of our capacity for color vision, which will be discussed in Chapter 6, evolved much more recently.

Recent discoveries concerning the genetic regulation of eye development have shed some light on the early evolution of eyes. Rebecca Quiring and her colleagues have shown that the homologous gene *Pax-6* controls the formation of the eye in fruit flies, mice, and humans. Remarkably, the mouse version of *Pax-6* remains sufficiently unchanged over evolutionary time that it can induce eye development in fruit flies. These findings suggest that *Pax-6* existed in the common ancestor of flies and mammals, an organism that existed before the great diversification of animals that occurred in the Cambrian period more than 500 million years ago. *Pax-6* possesses a homeobox DNA sequence and thus is related to the series of master regulatory genes. The existence of *Pax-6* suggests that the eyes of flies and humans, though varying greatly in structure, have a common evolutionary origin and thus are homologous. However, *Pax-6* is also found in nematodes, which do not even have photoreceptors, let alone eyes. In nematodes, this gene regulates the formation of the head, so it is perhaps more likely that the ancient role of *Pax-6* was to govern the shaping of the head and that it came to play a specialized role in eye formation separately in the lineages leading to flies and mammals. As the roles of *Pax-6* and other regulatory genes controlling eye formation become more completely understood, it may be possible to reconstruct the detailed history of eye evolution from the actions of genes in living animals.

## Eyes and Brain in a Chordate

We are members of the phylum chordates, distinguished by the presence during embryonic development of the notochord, a long, fibrous cord extending the length of the animal. In ancient free-swimming chordates and the living amphioxus, the notochord serves as a scaffold. In vertebrates the presence in the embryo of the notochord induces the development of adjacent structures in the nervous system. The nonvertebrate chordates have very simple nervous systems. The nonvertebrate chordate nervous system that has been best studied is from the larval amphioxus, which is free-swimming and lives by filtering micro-organisms from the water. In these animals there is a dorsal nerve cord running just above the notochord. The dorsal position of the nervous system in chordates contrasts with its ventral position in arthropods. It was this contrast in the location of the nervous system that led Geoffroy Saint-Hilaire in 1820 to propose that the ventral side of arthropods is homologous to the dorsal side of vertebrates, a suggestion that led to his contentious debate with Cuvier before the French Academy. It is now known from the work of E. DeRobertis and others that the genetic regulators of dorsal and ventral position are similarly inverted in arthropods and vertebrates, and that Geoffroy Saint-Hilaire was right.

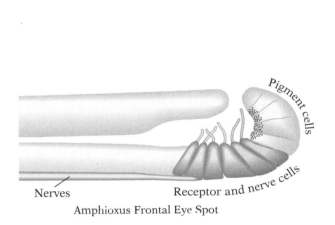

Amphioxus Frontal Eye Spot

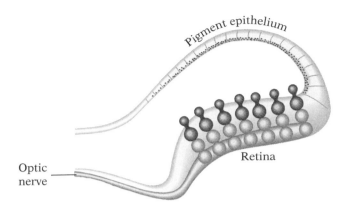

Developing Vertebrate Eye

The architecture of the nervous system in amphioxus may reveal the evolutionary origin of the vertebrate eye and parts of the brain. In an elegant study based on serial electron micrographs, Thurston C. Lacalli and his colleagues have shown the arrangement of structures in the amphioxus frontal eye spot located at the anterior end of the nerve cord in the cerebral vesicle. Pigment-containing cells are located in front; just below and behind them are ciliated cells that are probably photoreceptors, and neurons. This arrangement bears a topological correspondence to the embryonic vertebrate eye, in which the layer of pigmented cells overlies the photoreceptors, which are connected to neurons leading to the optic nerve. Thus the frontal eye spot in amphioxus appears to share a common plan with the far more complex pair of frontal eyes in vertebrates.

In the roof of the cerebral vesicle behind the eye spot is the lamellar body, a dense pile of photoreceptive cilia. The lamellar body may be homologous with the third or parietal eye found in many lower vertebrates and to the pineal gland in mammals. Like the parietal eye, the lamellar body may mediate daily activity cycles by responding to changes in sunlight throughout the day. (I will have more to say about the role of this system in regulating daily activity cycles in Chapter 7, in the section on brains and time.) In amphioxus the role of the lamellar body is probably to keep the larvae away from the surface waters during the day to avoid predation. Amphioxus larvae thus regulate their depth in the sea in a daily rhythm in response to sunlight.

Similarities between the structure of the frontal eye spot in amphioxus and the developing eye in vertebrates. The pigment cells overlying the eye spot in amphioxus may correspond to the pigment epithelium adjacent to the vertebrate photoreceptors.

Amphioxus

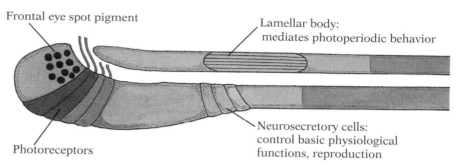

Frontal eye spot pigment

Lamellar body:
mediates photoperiodic behavior

Photoreceptors

Neurosecretory cells:
control basic physiological
functions, reproduction

Primitive Vertebrate

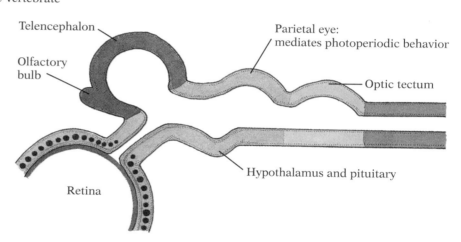

Telencephalon

Parietal eye:
mediates photoperiodic behavior

Olfactory
bulb

Optic tectum

Retina

Hypothalamus and pituitary

The brain of amphioxus compared with
that of a primitive vertebrate, based on
the work of Thurston C. Lacelli and his
colleagues.

In the floor of the cerebral vesicle are neurons that appear to correspond to the neurosecretory cells of the vertebrate hypothalamus,
which control reproductive functions, the timing of development,
and other basic aspects of physiology. Thus structures corresponding to three main components of the vertebrate forebrain (the
frontal eyes; the parietal eyes or pineal gland; and the neurosecretory system of the hypothalamus) are present in amphioxus, but
there is no evidence for another major component of the vertebrate
forebrain, the telencephalon. The telencephalon includes the brain
structures subserving olfaction and the cerebral cortex. On each side

of the brain a small fiber tract proceeds from the frontal eye to the midbrain, which is the major visual center in the vertebrate brain. There is evidence for the existence of a hindbrain in amphioxus based on the expression patterns of the homeobox genes and the location of serotonergic neurons in the dorsal nerve cord. These serotonergic neurons appear to correspond to the neurons in the vertebrate hindbrain described in Chapter 2. This remarkable constancy in the location of serotonergic neurons in amphioxus and vertebrates points to the fundamental nature of the serotonergic system and its stability in evolution. Taken together these findings reveal the presence in chordates of structures corresponding to parts of the vertebrate forebrain and hindbrain, while the telencephalon and midbrain appear to have originated in the earliest vertebrates.

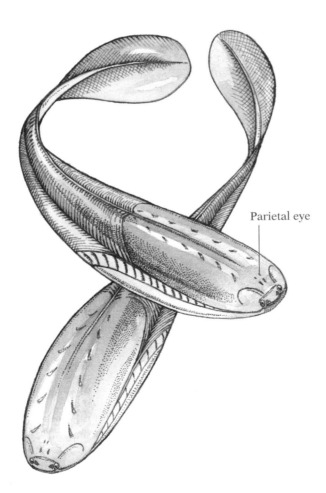

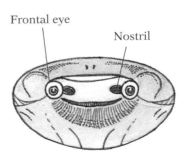

Sacabambaspis is one of the earliest known vertebrates. This fish lived in the ocean near the shore 450 million years ago.

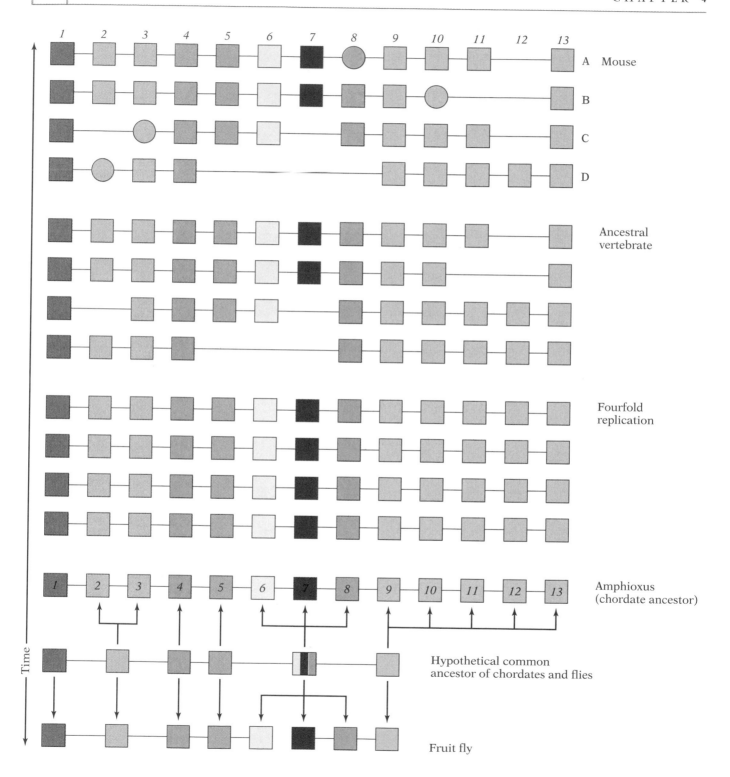

Opposite: The evolution of the homeobox genes in chordates and flies from an ancient common ancestor. The color-coding is the same as that in the diagram of mouse and fruit fly embryos on page 52. Note that some genes have been lost, and others, indicated by the circles, have been converted by mutations to "pseudogenes" and are not expressed. In fruit flies, the genes are located in two separate subsets, indicated by the lack of a connecting link between the yellow and red genes in the fly.

## The Rise of the Vertebrates

The earliest vertebrates, jawless fish, first appeared about 470 million years ago, shortly after the close of the Cambrian period. Above the mouth they had a pair of frontal eyes and nostrils for detecting prey and on top of the head a pair of parietal eyes that regulated their daily activity cycles. Their chordate ancestors had supported their modest life-style by filtering micro-organisms swimming in the water. In the early vertebrates, this way of life was replaced by a more vigorous existence that involved preying upon other organisms. Thus the earliest vertebrates, like the earliest amphibia, the earliest mammals, and the earliest primates, were small predators. Over and over again in evolution, the originators of new modes of life were small predators, and the key innovations at each stage conferred a selective advantage in predation.

Jordi Garcia-Fernandez and Peter Holland mapped the homeobox genes in amphioxus and found that they make up a single colinear set that contains all the individual genes homologous to the those found in the quadruple sets of homeobox genes in vertebrates. In each replicated set of vertebrate homeobox genes, some of the genes are missing. The primordial set would have had to contain the entire series of genes, and the homeobox series in amphioxus is exactly the set that would be predicted to have existed in the ancestor of vertebrates before replication into multiple sets with subsequent deletions. All vertebrates, including the primitive jawless fish, have multiple sets of homeobox genes, which suggests that at least one replication occurred about the time of the origin of vertebrates. There also is evidence that many other genes were replicated then. These multiple sets of master regulatory genes may have endowed vertebrates with greater combinatorial power in gene regulation and thus the capacity for greater differentiation of brain and body structures.

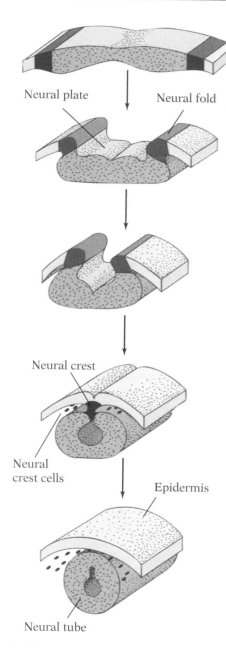

Neural plate

Neural fold

Neural crest

Neural
crest cells

Epidermis

Neural tube

The formation of the neural tube and the neural crest cells in the vertebrate embryo. The neural crest cells provide the plastic material for the evolution of new structures such as body armor, jaws, teeth, and the peripheral nervous system.

There is evidence that the control of the homeobox genes also changed at about the time of the origin of vertebrates. Ursula Dräger and her colleagues have noted that retinoic acid is made as a byproduct of photoreception in the eyes of vertebrates but not in invertebrates, and its role as a regulator of development also seems to be restricted to vertebrates. Dräger and her colleagues suggest that the role of retinoic acid in regulating gene transcription evolved in the eyes of early vertebrates and once established in the eye, the mechanism was taken over for the regulation of brain and limb development. An example, related in Chapter 3, is the stepwise activation of the homeobox genes by retinoic acid, which controls the formation of the hindbrain.

Another key innovation that occurred at about the time of the origin of vertebrates is the formation of the neural crest and its derivatives. The neural crest is a group of cells, unique to vertebrates, that originate in the lips of the neural groove in early embryos. As the neural groove closes to form the neural tube, the neural crest cells migrate away from the tube and become the precursors of the jaws, teeth, parts of the skull, and the peripheral nervous system. Like the neural tube, the formation of the neural crest is under the control of the homeobox genes. Many of the early vertebrates possessed body armor made of dentine plates that were neural crest derivatives, evidence of one of the early selective advantages conferred by this innovation. Thus in the early vertebrates the neural crest provided a reservoir of cells derived from the nervous system that were modified to perform new functions by homeobox genes, which gained new regulatory capacity by virtue of their replication into multiple sets.

As Carl Gans and Glenn Northcutt have pointed out, the new parts of the head in the early vertebrates included two sets of novel features that are related to the exchange and distribution of respiratory gases and to the detection and capture of prey. The greater metabolic demands of the more active life-style were supported by the gill apparatus and specialized muscles for respiration. The seizure of prey was enhanced by the development of pharyngeal muscles, and eventually by the formation of jaws in later fish. The detection of prey was enhanced by the development of the olfactory and visual systems along with their associated brain structures, the telencephalon and the optic tectum. The formation of the vestibular system and cerebellum developed in tandem with the visual system; their basic function is to support the stability of the retinal image in the eye during active movements by the animal.

Another key innovation that occurred early in vertebrate evolution was the fourfold replication of the gene that specifies the amino acid chain for primitive hemoglobin. In jawless fish, there is a single chain, but in all other vertebrates there are four similar chains that make up hemoglobin. Vernon Ingram showed that these four chains are the result of a fourfold replication of the original vertebrate hemoglobin gene. The four chains act cooperatively to bind and release oxygen more efficiently than does the single-chain variant. Since the brain is especially dependent on a reliable supply of oxygen, this change in the structure of hemoglobin may have facilitated brain evolution in jawed vertebrates. It is also an important reminder that replicated genes and structures often work in concert with one another rather than independently.

## Gene Duplications Create a Keen Sense of Smell

The early vertebrates were guided by a very keen sense of smell for the detection of chemicals in their watery environment. This new capacity was the result of a massive gene duplication process in which the genes for olfactory receptors were replicated over and over again. Linda Buck, Richard Axel, and their colleagues have found in living fish a family of olfactory receptor genes with about a hundred members. The olfactory receptor genes are related to the larger family of G protein–coupled receptor genes that includes those for the serotonin receptors described in Chapter 2. The various olfactory receptors bind to different odor-causing chemicals dissolved in water, and thus they provide an enormous capacity for discriminating olfactory stimuli. Smell provided a very sensitive mechanism not only for prey localization but also for forming memories to facilitate the future capture of nutritious prey and the avoidance of potentially toxic or otherwise dangerous prey. It also provided an important mechanism for social communication via olfactory signals, which would have facilitated reproduction. The olfactory receptors, which, like the photoreceptors, are modified cilia, line the passageway for the stream of water drawn through the head for the extraction of oxygen. These receptors have axons that project into the olfactory bulb at the front end of the brain. The olfactory receptor proteins may also serve as address codes that guide the axons of the receptor cells to their appropriate sites of termination in the brain. Bill Dreyer has proposed that these proteins may serve as developmental markers in other organs, such as the heart and testes. The olfactory bulb is part of the telencephalon, a new structure in vertebrates, although the precursors of the regulatory genes that control telencephalic development, such as the duplicated gene sets *BF-1* and *BF-2* and *Emx-1* and *Emx-2*, arose very early in evolution, as discussed in Chapter 3. Two of these ancient genes, *Emx-2* and *tailless*, have an important role in the formation of both the telencephalon and the olfactory sensory receptor organs.

The pioneering comparative neuroanatomists Ludwig Edinger and Cornelius Ariens Kappers, who worked in the early years of the twentieth century, proposed that the early evolution of the telencephalon was dominated by olfactory input and function. This "smell-brain" hypothesis was challenged by later neuroanatomists; however, the recent work by Helmut Wicht and Glenn Northcutt in the most primitive living vertebrates, the jawless hagfish, supports the smell-brain hypothesis for the origin of the telencephalon. The richly evocative capacity of odors to elicit memories of our past experience, particularly with respect to appetite and procreation, is the residue of this phylogenetically ancient development in the early vertebrates. The telencephalon of the hagfish and lampreys is smaller relative to body mass than in any other vertebrate group; nevertheless, this structure is complex even in these animals. Consequently, the telencephalon, the part of the brain that has undergone the greatest expansion in birds and mammals, began as a structure primarily devoted to the processing of olfactory information and the storage of olfactory memories.

## An Ancient Map

Unlike olfactory experience, which is not strongly linked to the geometrical space surrounding an individual, the visual space imaged on the retina has a high degree of topographic order. Our brains contain a storehouse of old maps, and one of the oldest of these is the representation of the retina on the roof of the midbrain, which is present in all vertebrates. This part of the midbrain is called the optic tectum, from the Latin word for "roof." The retina of each eye sends its axons to terminate in a topographic map in the optic tectum on the opposite side of the brain. The gene *BF-1* is expressed early in embryonic development in the nasal half of the retina in each eye. Its duplica, *BF-2*, is expressed in the temporal half of the retina in each eye. Junichi Yuasa and his colleagues demonstrated that these genes direct the formation of axonal connections between the retinae and the optic tecta and thus are in part responsible for the topographic order of the retinal maps in the optic tecta. Beneath the topographic map of the retina in the roof of the midbrain, there are inputs from the other senses, and thus the midbrain is an important center for the integration of spatial information from the different senses in all vertebrates.

It is not known why each retina projects to the optic tectum on the opposite side of the brain in vertebrates. In amphioxus and in the highly developed visual systems of cephalopods, the retinal fibers connect to the brain on the same side. Further studies of the genetic regulation of embryological development, it is hoped, will reveal the basis for the crossed pathways in vertebrates.

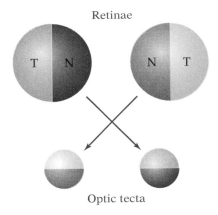

The mapping of the retina onto the optic tecta in vertebrates. N refers to the nasal half of each retina, the side toward the nose; T refers to the temporal half.

## The Origin of the Cerebellum

Successful predation requires rapid movements in the pursuit of prey. In order to see clearly while in motion, predators must have a mechanism for stabilizing the retinal image so that it will not be blurred by movement. Retinal stability is achieved through the vestibular system, which senses head movement and sends signals to the hindbrain. The hindbrain rapidly relays the signals to the extraocular eye muscles to move the eye in the opposite direction so as to compensate for head movements. In jawed vertebrates, movement of the head in each of the three planes of space is detected by the corresponding semicircular canal of the vestibular apparatus. Jawless hagfish, which lack the image-forming lens, also lack extraocular eye muscles and have only a very simple vestibular apparatus with a single semicircular canal. Lampreys and the fossil jawless fish have the two vertical semicircular canals that permit the correction for pitch and yaw, but lack the third, horizontal, canal found in jawed vertebrates.

A necessary part of this system for achieving a stable retinal image is the cerebellum, literally "little brain," which in the early stages of vertebrate evolution emerged from the roof of the hindbrain near where it joins the midbrain. The signals sent to the eye muscles must be precisely calibrated so as not to move the eye too

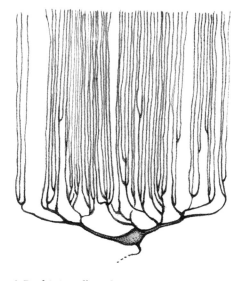

A Purkinje cell in the cerebellum of a mormyrid fish. The Purkinje cells integrate the inputs to the cerebellum. Note the geometric regularity of the dendrites, which may be related to the analysis of the precise timing of reflected electrical impulses.

rapidly or too slowly. Perhaps the most basic function of the cerebellum is to compare eye velocity with head velocity and adjust the signal sent to the eye muscles so that the retinal image is stabilized. The cerebellum is poorly developed or absent in hagfish but is present in lampreys and in the fossil jawless fish. It has expanded enormously in mormyrid fish, where its functions may be related to the need for precise coordination between the emission of electric pulses and the reception of reflected signals by electroreceptors distributed across the surface of the fish's body. In the cerebellum of mormyrids, the arrangement of the dendrites of neurons is an almost crystal-like lattice, which may support precise electric timing in this circuitry.

## Myelin: A Crucial Vertebrate Innovation

Myelin is the material that insulates axons; it is fundamental to the functioning of the brain in higher vertebrates. Myelin is made of protein and fat molecules that are formed into sheets by specialized cells, the oligodendroglia, that wrap around axons. Myelin insulates each axon so as to create "private lines" that are not contaminated by cross talk from other nearby axons. It also greatly increases the speed and energy efficiency of axonal conduction. This is achieved through a unique mechanism called saltatory ("jumping") conduction. The myelin insulation restricts the flow of ions in and out of the axon, but there are gaps in the myelin spaced at about 1-millimeter intervals along the axon. The flow of ions that restores the action potential as it moves down the axon can occur only at the gaps, and thus the action potential jumps from one gap to the next. The speed at which axons can conduct action potentials increases with their diameter. Because of their insulation, myelinated axons conduct much more rapidly than do unmyelinated axons of the same diameter, with the result that many more myelinated axons can be packed into a limited volume of space. Thus myelinated axons can support a more richly interconnected brain. Saltatory conduction also increases efficiency because the expenditure of energy to restore the balance of ions after the passage of an action potential is needed only at the gaps and not all along the axon as is the case for unmyelinated axons.

After exhaustive searches by Theodore Bullock and others, myelin has not been found in any invertebrate or in the jawless verte-

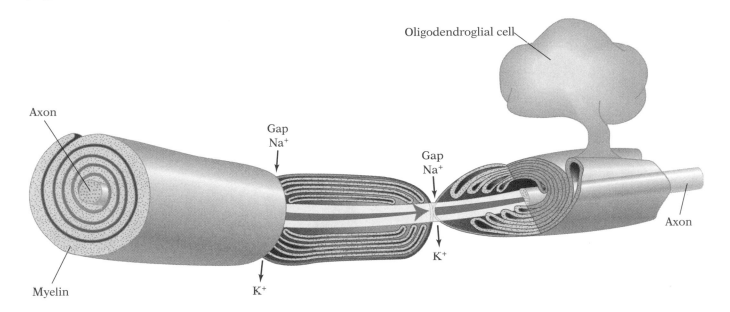

brates, the hagfish and lampreys. However, oligondendroglia and myelinated axons are present in all jawed vertebrates. The innovation of myelin may have enabled the early jawed vertebrates to assume more active predatory behavior as compared with the scavenging or parasitic mode of life characteristic of hagfish and lampreys. John Gerhart and Marc Kirschner point out that all the molecular components of myelin are present in other animals, but it appears to be the innovation of the myelin-making cells, the oligodendroglia, that is responsible for this evolutionary advance in jawed vertebrates. As with other evolutionary advances, myelin carries with it vulnerability to disease, in this case the risk of its degeneration, which occurs in the common and devastating condition multiple sclerosis. In this disease axonal conduction and thus the ability of neurons to communicate is grossly compromised.

## Cephalopods: The Second Great Pinnacle of Brain Evolution

The other group of animals that have evolved large eyes and brains are the cephalopods: the nautilus, squid, octopus, and cuttlefish. In terms of relative brain size, some of the larger-brained cephalopods,

The structure of myelin insulation and the saltatory conduction of action potentials in the brains of jawed vertebrates. The oligodendroglia extrude membrane that forms the myelin sheath that wraps around axons. The action potential jumps from one gap in the myelin to the next. The action potential is renewed at these gaps by the influx of sodium ions ($Na^+$) into the axon and the efflux of potassium ions ($K^+$) from the axon. April Davis and her colleagues have recently discovered myelinated axons in another group of animals, the copepod crustaceans. Copepods are among the most abundant animals living in the world's oceans. Their great success may be due to their ability to escape rapidly from predators, and this ability may arise from the fast conduction of nerve impulses by their myelinated axons. As in vertebrate myelinated axons, there are periodic gaps in copepod myelinated axons to allow the passage of ions as action potentials jump along the axons.

Left: A dissected nautilus drawn by Richard Owen. Half the shell has been removed to reveal the chambers.
Right: A living nautilus devouring a piece of fish.

the octopus and cuttlefish, fall within the lower part of the mammalian range. The evolution of the sensory systems and brain in cephalopods show remarkable parallels with vertebrates, although their common ancestor must have been a very primitive animal possessing an extremely simple brain and photoreceptors. Primitive cephalopods first appeared in the late Cambrian period, about 500 million years ago. The earliest cephalopods resembled the chambered nautilus that lives today in the depths of the Indian Ocean. Like the early vertebrates, the cephalopods were predators, and like them they developed specialized respiratory mechanisms to support a more active mode of life and more elaborate sensory and motor mechanisms for the detection and capture of prey. Like the early vertebrates, the nautilus has a well-developed olfactory system. It also has a primitive eye, which lacks a lens and instead has a simple pinhole aperture. The nautilus also has a simple organ, the statocyst, for detecting body movement, and has some capacity for adjusting its eye in response to body movement and thus achieving retinal stability.

Five hundred million years ago, cephalopods invented jet propulsion. They move by rapidly expelling seawater from their body cavity with a sudden powerful muscular contraction that is triggered

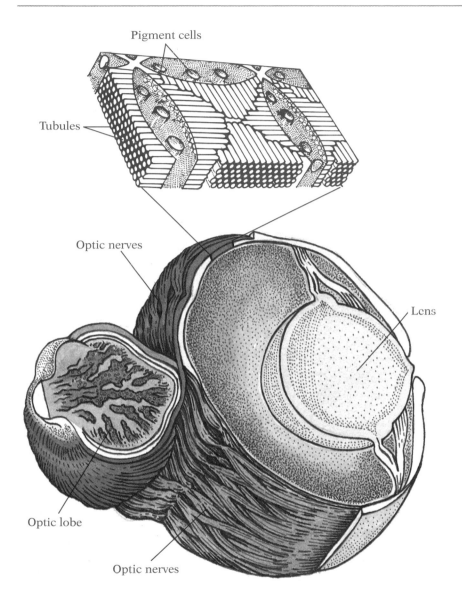

Pigment cells

Tubules

Optic nerves

Lens

Optic lobe

Optic nerves

The structure of the octopus eye, based on the studies of J. Z. Young. The visual image is inverted by the lens. The neural representation of the retinal image is reinverted by the dorsoventral crossing of the myriad optic nerves, which project topographically onto the optic lobe. The upper diagram illustrates a section through the retinal photoreceptors. The photoreceptive membranes are located in the microvilli, which are oriented in the horizontal or vertical plane, and which probably analyze the plane of polarized light. The axons of the photoreceptors form the optic nerves. Thus the octopus eye projects directly to the brain, unlike the vertebrate retinal photoreceptors, which connect to a series of neurons within the retina itself before connecting via the optic nerve to the brain.

by a cascade of impulses from giant axons. The more advanced cephalopods, the octopus, squid, and cuttlefish, have magnificently developed eyes related to their active predatory life-styles. The photoreceptors are organized into a precise horizontal–vertical lattice that enables cephalopods to discriminate the plane of polarized light, a capacity lacking in vertebrates. In the octopus there is a

The octopus brain viewed from above. The large kidney-shaped structures are the optic lobes for each eye, which contain most of the neurons in the brain. The optic lobes connect to the vertical lobes, which have memory functions; below these lobes are the centers for the control of the arms and mouth.

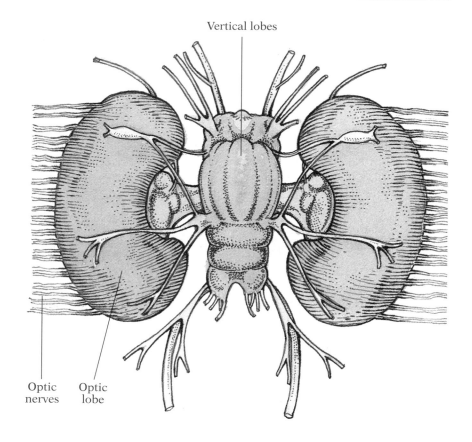

Vertical lobes

Optic nerves

Optic lobe

vertical crossing of the optic fibers so that they map onto the optic lobe in such a way as to the re-invert the virtual image that had been inverted by the lens. Thus the representation of the retinal image in the optic lobe is returned to its original upright position. The advanced cephalopods have also developed in their statocysts a set of detectors for movement in the three planes of space in a manner remarkably analogous to the three semicircular canals found in jawed vertebrates. There is even a brain structure analogous to the cerebellum for regulating amplification in the feedback loop between the movement receptors and the extraocular eye muscles for achieving retinal stability. Finally, in the advanced cephalopods there is a specialized system for visual memory storage located in the vertical lobe.

The evolution of the brain in cephalopods was fundamentally limited by their failure to develop a cell type analogous to oligodendroglia to manufacture an insulating material like myelin. Thus more space and energy is taken up by axons in cephalopods than in jawed vertebrates. Another serious constraint on brain evolution in cephalopods is the oxygen-carrying capacity of their vascular system. Cephalopod blood—which is green—contains hemocyanin, a copper-based protein, which transports oxygen to the tissues. Hemocyanin can carry only about one quarter as much oxygen as the iron-based hemoglobin in vertebrates. Thus vertebrate brains have much more oxygen available to support their activity. These comparisons illustrate that a small number of biophysical mechanisms have had an enormous impact on the course of brain evolution. In the next chapter I will explore how another set of biophysical mechanisms, those for maintaining a constant body temperature, influenced the evolution of the brain in mammals and birds.

A juvenile Japanese macaque monkey making a snowball. The scene illustrates two basic features of mammals: most mammals can maintain a constant body temperature across a broad range of environmental temperatures, and all young mammals engage in activities with no direct payoff in terms of enhanced survival, behaviors that are easily recognized as play. The indirect benefit of play is that it facilitates the maturation of the cortical systems of the brain.

# Warm-Blooded Brains

No single characteristic could evolve very far towards
the mammalian condition unless it was accompanied by
appropriate progression of all the other characteristics.
However, the likelihood of simultaneous change in all
the systems is infinitesimally small. Therefore only a
small advance in any one system could occur, after which
that system would have to await the accumulation of
small changes in all the other systems, before evolving a
further step toward the mammalian condition.

T. S. Kemp,
*Mammal-like Reptiles and the Origin of Mammals,* 1982

The brains of warm-blooded vertebrates, the mammals and birds, tend to be larger than the brains of cold-blooded vertebrates of the same body weight. The larger brains in mammals and birds are a crucial part of a large set of mechanisms for maintaining a constant body temperature. Since all chemical reactions are temperature dependent, a constant body temperature brings about stability in chemical reactions and the capacity for the precise regulation and coordination of complex chemical systems. However, maintaining a constant body temperature requires a tenfold increase in energy expenditure. The great increase in energy metabolism puts enormous demands on the sensory, cognitive, and memory capacities of the brain in warm-blooded vertebrates because they must find much larger amounts of food than cold-blooded animals. Why was it advantageous for them to pay such a huge price, and what changes occurred in the brain?

## The Invasion of the Land: Predators Lead the Way

The first amphibians, the distant ancestors of mammals and birds, crawled out of the water about 370 million years ago. On land they encountered greater and more rapid fluctuations in environmental temperature than their fish forebears had in the water. They developed a second set of olfactory receptors located in the roof of the mouth. This set, called the vomeronasal system, is connected by a separate pathway to the brain. This system appears to be particularly involved in the detection of pheromones, chemical messengers that serve to communicate sexual receptivity and other information to members of the same species. Recently two teams, Hiroaki Matsunami and Linda Buck and Gilles Herrada and Catherine Dulac, have discovered a new gene family for this set of olfactory receptors that is distinct from the family of receptor genes expressed in the main olfactory bulb. As with the main olfactory gene family, this second large family may also serve as developmental markers in other organs.

About 70 million years after the first amphibians crawled ashore, the first reptiles appeared. Their innovation was eggs that could be laid on land. The eggs were enclosed in a semipermeable shell that allowed the embryo to breathe but protected it from drying out. The

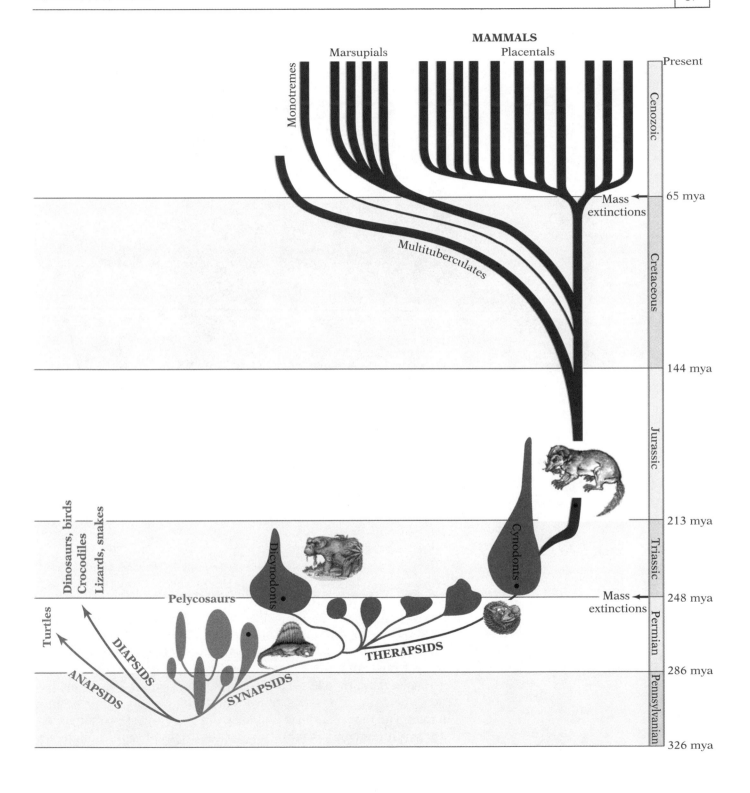

**MAMMALS**
Marsupials            Placentals

Monotremes

Multituberculates

Mass
extinctions ← 65 mya

Present

Cenozoic

Cretaceous

144 mya

Jurassic

213 mya

Cynodonts

Dinosaurs, birds
Crocodiles
Lizards, snakes

Triassic

Pelycosaurs

Dicynodonts

Mass
extinctions ← 248 mya

Turtles

DIAPSIDS

THERAPSIDS

Permian

286 mya

ANAPSIDS

SYNAPSIDS

Pennsylvanian

326 mya

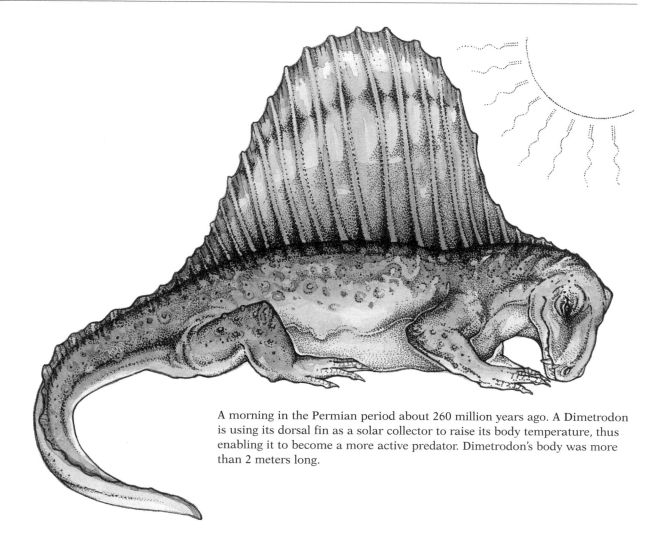

A morning in the Permian period about 260 million years ago. A Dimetrodon is using its dorsal fin as a solar collector to raise its body temperature, thus enabling it to become a more active predator. Dimetrodon's body was more than 2 meters long.

embryo was also protected by the amniotic membranes, which maintained the appropriate fluid environment and enclosed the food supply. Thus the reptiles did not need to be close to a body of water and could range farther inland, thus exposing themselves to an even broader range of environmental conditions.

The earliest reptiles were small predators about 20 centimeters long, the first of a series of vertebrate innovators to prey upon insects. The first mammals and probably the earliest primates were also insect predators. Very soon after the origin of reptiles there was a basic division into three separate lineages: the synapsids, which led to mammals; the anapsids, which led to the living turtles; and the diapsids, which led to dinosaurs, birds, and other living reptiles apart from turtles. The basic distinctions among these three great lineages are related to the shape of the skull and the arrangement of the muscles that close the jaw, both specializations related to predatory behavior. The earliest members of the line leading to mammals were the pelycosaurs. The early pelycosaurs were also small predators, about 60 centimeters long. The new feature in the pelycosaurs was the development of a more efficient chewing mechanism and daggerlike canine teeth with which to stab their prey. This change is the beginning of the progressive specialization of the teeth that is a hallmark of mammals and stands in strong contrast to reptilian teeth, which all have the same structure. A very successful pelycosaur was Dimetrodon, which had a huge sail rising from its vertebrae. The sail was an early experiment in temperature regulation. It served as a solar collector that boosted Dimetrodon's metabolic activity in the morning, thus enabling it to gain a selective advantage over its more sluggish contemporaries. The sail may also have served as a radiator to cast off excess body heat during periods of intense exertion. In our time, the long deceased Dimetrodon continues to stir the imagination—it was the subject of a prize-winning float in the Rose Bowl Parade in Pasadena, California, in 1993!

It is conventional to think of a series of fossils progressing in linear succession down to a modern descendant. In fact, the pelycosaurs were an example of broad adaptive radiation derived from a common ancestor, which produced a wide variety of animals that invaded a great diversity of ecological niches. Most of these became evolutionary dead-ends within 50 million years. However, the pelycosaurs gave rise to a new group, the therapsids, of which the earliest members were again small predators with very elongated canine teeth. The therapsids, like their pelycosaur predecessors, underwent another huge adaptive radiation, which produced gigantic herbivorous forms as well as predators. They were the dominant land animals of their time, but they were nearly all wiped out at the end of the Permian period, 248 million years ago in the most catastrophic mass extinction in the earth's history. Near the end of the Permian period, there were a series of massive volcanic eruptions in Siberia that produced a flow of basalt that covered 600,000 square miles. The dust and gases arising from these massive eruptions may have caused a global cooling at the earth's surface and thus brought

A morning in the early Triassic period, about 245 million years ago; these animals survived the mass extinctions at the end of the Permian period. The lumbering giant herbivore is the dicynodont Lystrosaurus, which was extremely abundant in the early Triassic but which became extinct a few million years later. The much smaller, possibly furry pair of animals are Thrinaxadons, which were cynodonts, members of the group that gave rise to mammals. The small reptile is a thecodont, also a predator, and a member of another group with a great future giving rise to the dinosaurs and birds. The cutting edge of evolution often occurs in small predators.

about the extinctions, in which an estimated 95 percent of all animal species died out.

Among the few tetrapods to survive the Permian extinctions was a giant herbivorous therapsid, Lystrosaurus, that was extremely abundant during the early years of the following Triassic period. There were also two predators present in this impoverished fauna. One was a thecodont, which gave rise to the birds and the dinosaurs, which displaced the large therapsids as the dominant land animals. The other was a small therapsid, a cynodont, so named because of its doglike canine teeth, which gave rise to the mammals.

## Staying Warm and Keeping Cool

Most mammals and birds live at a relatively constant body temperature. The maintenance of a constant body temperature—temperature homeostasis—is a very expensive process. Resting mammals and birds typically expend 5 to 10 times as much energy as do comparably sized reptiles, and the great bulk of this increased energy expenditure is devoted to maintaining temperature homeostasis through muscular exertions to heat and cool the body. This commitment to temperature homeostasis means that mammals and birds must find about an order of magnitude more food to eat than reptiles of the same size.

The rates of virtually all chemical reactions in living systems are temperature dependent. Most life processes involve a series of biochemical reactions each of which is dependent on the preceding steps and may be influenced by feedback from subsequent steps in the series. If the reactions at different steps proceed too fast or too slowly, the whole process will be compromised. The regulation of these highly interdependent reactions can be more efficient if they run at a constant temperature rather than over a widely varying range of temperatures. So long as the benefits of more efficient biochemical regulation exceed the energy costs of maintaining a constant body temperature, there is a selective advantage for temperature homeostasis. This is not an all-or-none process. Homeostasis is a buffer that protects the organism from changes in environmental temperature. With increased expenditures the buffer can offer more protection for larger temperature variations or for longer periods of time. Both increased energy from food and changes in the brain, body, and behavior are required to support improved homeostasis. These changes are crucial features of the evolutionary history of mammals and birds over the past few hundred million years. They involve changes in the quantity of food consumed and the way it is chewed, in breathing, in locomotion, in parenting behavior, in the senses, in memory, and in the expansion of the forebrain.

In the mammalian line the anatomical and physiological changes responsible for temperature homeostasis occurred in stages. In the earliest members of the line leading to mammals, the parietal eye was well developed and probably partipated via connections with the brain in temperature-regulating behavior related to the daily cycle of changes in light and temperature; this is the function of the parietal eye in living reptiles. Some early members of this line, the pelycosaurs, experimented with a novel means for thermoregulation, the dorsal sail. Later members of the line leading to mammals, the cynodonts, evolved many features indicative of a more active lifestyle. The first mammals were most likely nocturnal and were thus inactive during the higher daytime temperatures. The ability to sense the daily light cycle directly by the parietal eye was lost in the early mammals, but the daily changes in light were relayed to the pineal gland via an indirect route from the main eyes. (This circuit is the brain's clock and I will have more to say about it in Chapter 7.) Restriction of activity to the cooler nighttime would have facilitated the development of temperature homeostasis in the early mammals, since it is less energetically expensive to heat the body than it is to cool it. Shivering and other forms of muscular activity can readily generate heat, but the only mechanism available for reducing body temperature is evaporative cooling obtained through panting and sweating. The early mammals probably maintained their body temperature at a level just a little above the nighttime temperature. Eventually some mammals became active during the day and thus were exposed to the higher daytime temperatures. Today, mammals maintain their body temperature at about 37 to 39 degrees Celsius, which is near the high end of the range of the daytime temperatures to which they are exposed, because it is easier to heat the body than to cool it.

The body temperature and resting metabolic rate are even higher in birds than in mammals, but the fossil record is much less abundant. It is not clear how or when the transition was made from cold-blooded reptile to warm-blooded bird, although we do know that some of the early birds had feathers that served as insulation.

# The Cynodonts Become More Active

The cynodonts were predators that ranged in size from that of modern ferrets to wolves. Very successful animals in their time, they are known from many well-preserved fossils. They had many innovations that were related to a more energetic life-style. Cynodont teeth were differentiated into incisors, canines, and molars for the cutting, piercing, and grinding of food, preparing it for the more rapid digestion necessary to support a higher metabolic rate. The molar teeth developed multiple cusps for grinding the food. The teeth are derived from neural crest cells that have migrated away from the developing hindbrain, and thus in a sense teeth are displaced and transformed bits of brain tissue. Recently, Bethan Thomas and her colleagues have found that the formation of the molar teeth is under the control of *Dlx-1* and *Dlx-2*, a duplicated pair of genes containing the homeobox sequence and thus related to the homeotic series. *Dlx-1* is also expressed in the ventral parts of the developing forebrain. Thus the differentiation of the teeth for specialized functions in preparing food for rapid digestion and the formation of the forebrain are under the control of the same gene, and this implies a close linkage in the evolution of the teeth and forebrain in the mammalian line.

The cynodonts are believed to have had a muscular snout and lips, a conclusion based on the presence of well-developed passages through the skull for the transit of nerves and blood vessels supplying these structures. Muscular lips would have enabled baby cynodonts to suckle their mother's milk. The snout may have been covered by a system of whiskers for the tactile perception of food objects. The jaw muscles were huge. There also were important changes in the cynodont respiratory system. The bony palate formed a barrier between the oral and nasal cavities so that the cynodonts could swallow and breathe at the same time, a necessary adaptation to prevent choking in animals with a high metabolic rate. Within the nasal cavity a complex set of thin bones called the turbinals bore the olfactory membranes, which warmed and humidified the incoming airstream for more efficient respiration. The expanded olfactory membranes in cynodonts also had a larger surface area for olfactory chemoreception, leading to a heightened sense of smell in these animals. This was probably the time of the huge expansion of the family of olfactory receptor protein genes through a wave of massive gene duplications that created the mammalian set of about 1000 genes. Fossil endocasts of cynodont brains indicate that the

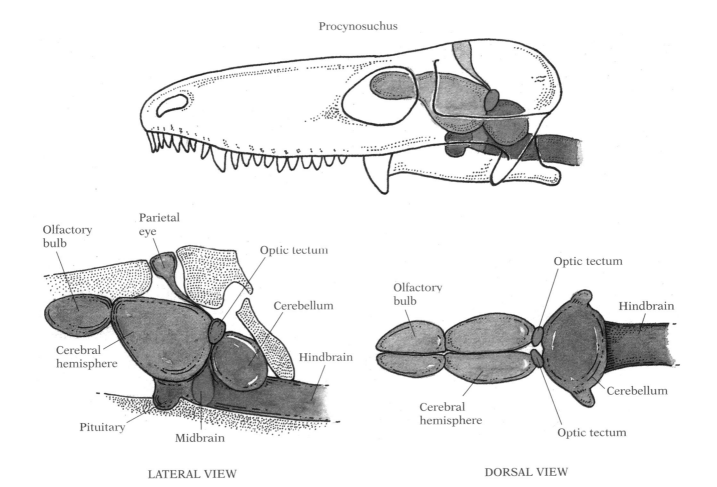

Procynosuchus

**Olfactory bulb**

**Parietal eye**

**Optic tectum**

**Cerebellum**

**Cerebral hemisphere**

**Hindbrain**

**Pituitary**

**Midbrain**

LATERAL VIEW

**Olfactory bulb**

**Optic tectum**

**Hindbrain**

**Cerebral hemisphere**

**Cerebellum**

**Optic tectum**

DORSAL VIEW

olfactory bulb was large. Overall the brains of cynodonts were intermediate in size between living reptiles and mammals of comparable body size. Through the action of homeotic genes, there was a reduction and eventual loss of ribs on the lumbar vertebrae, and a muscular diaphragm formed to separate the thoracic and abdominal cavities. These are specializations associated with active respiration in mammals. Finally, we know that cynodonts slept in a curled-up, energy-conserving posture, much the way mammals do today.

A reconstruction by T. S. Kemp of the brain of Procynosuchus, an early cynodont. The upper diagram shows the brain in its location within the skull.

A cynodont, Thrinaxadon, fossilized in sleeping posture and discovered 240 million years later by A. S. Brink.

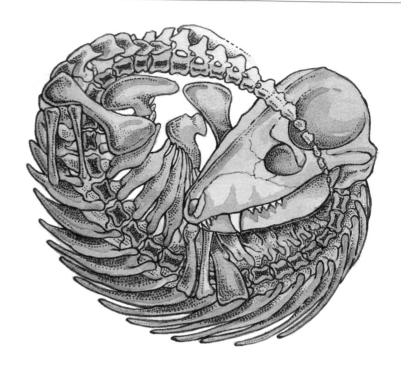

The evidence for a higher metabolic rate suggests that the cynodonts had begun to develop parenting behavior, which is inextricably linked to temperature homeostasis. Parenting behavior occurs in some cold-blooded animals, but it is universal among warm-blooded vertebrates. Temperature homeostats must devote most of their energy to maintaining their body temperature, a need that conflicts with the nutritional burden of the growth process. Thus, during the early stages of postnatal development baby homeostats must obtain their food and warmth from an outside source. Mammals have responded to this requirement of homeostasis with the formation of mammary glands and lactation in females, which of course is the original defining characteristic of the class Mammalia. The mammary glands are specialized sweat glands, indicating that they derive from a more ancient system for the evaporative cooling of the body following exertion. Milk is a complex food containing more than 40 nutrients, including sugars, fats, and proteins. The functional maturation of the brain probably depends on this precise mix of nutrients, since there is evidence that human infants

## The Superimposition of Maps in the Tectum

Ursula Dräger and David Hubel found that the visual and tactile maps are superimposed in the midbrain of the mouse. This convergence of visual and tactile information serves to orient the mouse to novel stimuli. Such a convergence is likely to have occurred in the midbrain at the time when whiskers first appeared in the cynodonts.

There are many other examples of the superimposition of spatial information from different senses in the midbrains of vertebrates. One of the most striking was revealed by the work of Eric Knudsen in owls. He found a spatial map of auditory space that is aligned with the visual map in the optic tectum. This superimposition of visual and auditory spatial information serves to facilitate prey localization by the owl.

raised on natural human milk have significantly higher IQs than infants raised on formula milk when both are administered through bottles.

The ejection of milk from the mammary gland is under the control of the hormone oxytocin, which is made in the hypothalamus in the basal part of the forebrain. Oxytocin also stimulates maternal care in mammals, and it is likely that it came to have this function in the cynodonts. Oxytocin in a member of an ancient family of hormones that controls reproductive and other physiological functions in both vertebrates and molluscs, and its expanded role in the mammalian line illustrates how old components of the nervous system can assume new functions. In mammals the huge energetic burden of sustaining the growth of infants falls on the mother. In small mammals, lactation *triples* the amount of food that must be eaten by a female. There is some direct evidence of parenting behavior in cynodonts. In 1955, the paleontologist A. S. Brink found the fossil of a baby cynodont nestled next to a much larger individual, and the pair may be an infant with its mother.

## The First Mammals

Toward the end of the Triassic period, about 220 million years ago, the first true mammals appeared. They were very much smaller than their cynodont ancestors, which weighed more than a kilogram. The first mammals weighed less than 30 grams and resembled the shrews that live today. They were very active predators with major innovations in the brain, in hearing mechanisms, and in tooth development. In contrast to these progressive changes, the visual system was reduced in the early mammals. The cynodonts had large eyes that were protected by a bony pillar called the post-orbital bar, but in the early mammals the eyes lost this protection. This lack of eye protection is characteristic of small-eyed nocturnal mammals living today like the shrews, hedgehogs, tenrecs, and opossums. In fact, the great majority of living mammalian species are mainly active at night, further evidence of an ancient mammalian heritage of nocturnality. Another inference that can be made about the behavior of the early mammals from living shrews and

Megazostrodon, one of the earliest mammals, lived in the late Triassic period. It is shown here slightly larger than life size. It is another example of an evolutionary advance made by a small predator. Megazostrodon's small size relative to its cynodont ancestor is shown by the comparison in the inset.

opossums is that they had a very simple social structure in which adults were solitary except for nursing mothers, which carried the full responsibility for rearing their offspring.

There was a major transformation of the hearing apparatus in the earliest mammals. Two bones that were part of the jaw joint in the cynodonts became incorporated into the hearing apparatus of the earliest mammals to form the chain of ossicles that conducts sound from the eardrum to the inner ear. These two bones, the articular and the quadrate of the cynodont jaw, became the malleus and the incus of the mammalian ear. The third member of the mammalian chain of ossicles, the stapes, was already serving as the conductor of sound and is still the sole conductor in amphibians, reptiles, and birds. This amazing transformation of jaw bones into the ossicles of the middle ear was first observed in developing pig embryos by C. B. Reichart in 1837 and subsequently found in the fossils of the earliest mammals.

The functional advantage of the chain of ossicles appears to be related to the capacity of mammals to discriminate much higher frequencies than reptiles and birds are able to hear. Hearing in non-mammals is limited to less than 10,000 cycles, whereas mammals can hear much higher frequencies, sometimes above 100,000 cycles. In mammals the stapedius muscle adjusts the stiffness of the linkage

A living nocturnal insectivorous mammal, the tenrec, *Hemicentetes sembrinosus*. Tenrecs mature very rapidly and have short life spans.

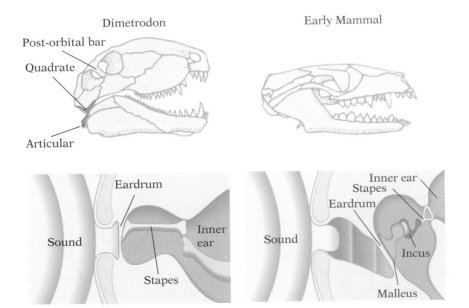

The evolution of jaw bones into the ossicles of the middle ear in mammals. In Dimetrodon the articular and quadrate bones formed part of the jaw joint; the stapes conducted sound from the eardrum to the sound receptors in the inner ear. In mammals, the articular was transformed into the malleus and the quadrate into the incus of the middle ear. In mammals, the malleus and incus, together with the stapes, make up the chain of ossicles that transmits sound from the eardrum to the inner ear, which was the site of another major mammalian innovation, the outer hair cells. Note also that in Dimetrodon a bony strut called the post-orbital bar protected the eye. The post-orbital bar disappeared in early mammals, indicating the loss of importance of vision in these animals.

The frequency range of hearing in amphibians, reptiles, birds, and mammals plotted in terms of hearing thresholds in decibels. Mammals hear much higher frequencies than do other vertebrates; note that the sound frequencies are plotted on a logarithmic scale. These curves are based on audiograms for a large number of species compiled by Richard R. Fay. They represent the outer envelope of all the curves for each class of vertebrates.

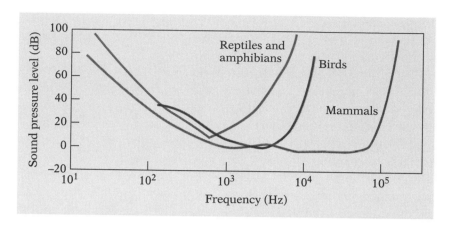

The stereocilia of the inner hair cells (above) and the outer hair cells (below) in the inner ear of the platypus, a monotreme, imaged by A. Ladhams and J. Pickles. Outer hair cells are a unique mammalian feature and are present in all mammals.

between the ossicles. When the stapedius contracts it reduces the transmission of low-frequency sounds, thus enabling the hair cell receptors in the cochlea to resolve high-frequency sounds.

The auditory physiologist William Brownell has proposed that the capacity to hear at higher frequencies is also related to another uniquely mammalian feature, the outer hair cells of the cochlea. Cochlear inner hair cells resemble hair cells in all nonmammalian inner ears both in terms of their structure and in their ability to analyze the acoustic spectrum. The capacity of inner hair cells to analyze higher frequencies is linked to the functioning of the outer hair cells, which are arranged in three concentric rows parallel to the row of inner hair cells in the cochlear spiral. A remarkable discovery made initially by Brownell and his colleagues in 1985, and followed up by a number of other researchers, is that the outer hair cells can change their shape extremely rapidly in response to sound and that this mechanical change leads to an enhanced capacity of the inner hair cells to discriminate the higher frequencies. Thus the rows of outer hair cells are another example of duplicated structures that act cooperatively with the original structure to enhance functional capabilities during the course of evolution.

The capacity to hear higher frequencies was very advantageous to the early mammals, enabling them to resolve the high-frequency noises made by their insect prey and thus facilitating their capture. This capacity also enabled them to detect high-frequency distress sounds made by their own infants. All baby mammals cry when cold, hungry, or separated. In small mammals these cries are typically

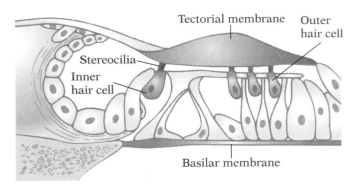

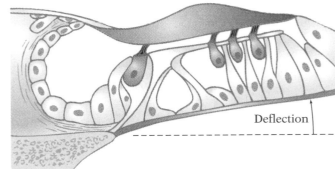

at very high frequencies; for example, distressed baby mice cry at about 25,000 cycles. Eyo Okon showed that cooling elicited ultrasonic cries in baby mice. Babies that have wandered from the nest also make ultrasonic cries, which cause their mothers to retrieve them. These high-frequency cries and the capacity to hear them provided the early mammals with a private channel of communication between baby and mother that was inaudible to reptilian and avian predators. Thus the evolution of the capacity for high-frequency hearing manifest in the transformation of jaw bones into ear ossicles and the innovation of the outer hair cells were closely linked to the development of parental care, which in turn was linked to the acquisition of temperature homeostasis.

The second great mammalian innovation was in the way the teeth developed. The cynodonts, like reptiles, grew slowly, and their teeth were continually replaced throughout life as they wore out. The early mammals were essentially miniature cynodonts that matured when they were still the size of cynodont infants. They had during the course of their lives only two sets of teeth, the deciduous and permanent teeth, as in most modern mammals. The presence of a single set of permanent teeth in adult life permitted a more precise fit between the cusp surfaces of the upper and lower molars, thus providing more efficient chewing, which in turn would have facilitated the more rapid digestion of finely ground food and thus a higher metabolic rate. This condition contrasts with that of the cynodonts, in which the teeth were continually being replaced and the upper and lower teeth were mismatched.

The inner and outer hair cells are in contact with the tectorial membrane. Sound vibrations cause the basilar membrane to move, in turn causing the stereocilia of the hair cells to bend. The outer hairs are themselves motile and influence the mechanical response of the inner hair cells, enhancing their capacity to respond to high sound frequencies. Only the inner hair cells relay auditory input to the brain; the motile response of the outer hair cells is controlled by a feedback pathway from the brain.

A scene from the late Triassic, about 200 million years ago. A baby mammal has strayed from its nest and is making a high-frequency cry that is audible to its mother (but not to a nearby predator) and alerts her to her infant's peril. The clueless reptile is unaware of the baby's distress because its cries are well above the reptile's hearing range.

Mammalian development is thus a truncated version of cynodont development. One consequence of this truncated development was that the early mammals were able to specialize on insect prey, whereas large reptiles, like crocodiles, must shift to larger prey as they themselves grow larger. Stephen Jay Gould has emphasized that mutations that affect the timing of development appear to be important mechanisms underlying many major transformations throughout evolutionary history. Over time, the acceleration and truncation of development result in a descendant group in which the adults resemble the young of their ancestors. This process is called pedomorphism, from Greek words meaning "child shaped." Because the proportion of brain size to body size is greater in infants than in adults, pedomorphic changes result in increases in relative brain size. The existence of a well-defined developmental cycle of accelerated maturation reaching an adult plateau contrasted with the gradual reptilian mode of development in the cynodonts, in which there

## Is Senescence Adaptive?

Caleb Finch has suggested the intriguing possibility that the changes in the developmental program in early mammals also included declining function in adulthood, or senescence. Finch, in his massive comparative study of aging, *Longevity, Senescence, and the Genome*, found that different groups of animals vary greatly in the mode and tempo of senescence. Many reptiles exhibit very slow or even negligible senescence. Negligible senescence is especially well documented for natural populations of turtles in the careful, long-term studies by Justin Congdon and his associates. In the late 1870s, Alfred Russel Wallace, the co-conceiver of the theory of natural selection, suggested that senescence was a evolutionary adaptation to reduce competition that would otherwise develop between succeeding generations. Wallace reasoned: "If individuals did not die, they would soon multiply inordinately and would interfere with each other's healthy existence. Food would become scarce, and hence [if] the larger individuals did not die they would decompose [starve]. The smaller organisms would have a better chance of finding food, the larger ones less chance. That one which gave off several small portions to form each a new organism would have a better chance of leaving descendants like itself than one which divided equally or gave off a large part of itself. Hence it would happen that those which gave off very small portions would probably soon after cease to maintain their own existence while they would leave numerous offspring. This state of things would therefore by natural selection soon become established as the regular course of things, and thus we have the origin of old age, decay, and death; for it is evident that when one or more individuals have provided a sufficient number of successors they themselves, as consumers of nourishment, are an injury to those successors."

Wallace created a model system to explain the evolution of senescence. His early model predicted what subsequent research suggests happened in the evolution of the cynodonts into mammals. The cynodonts were large, probably slowly developing, and long-lived judging by their continuous replacement of worn teeth. Their descendants, the first mammals, were small, probably rapidly developing and short lived as is the case for living marsupial and insectivorous mammals that most closely resemble them. The higher metabolic rate associated with improved temperature

Alfred Russel Wallace (1823–1913), coconceiver of the theory of natural selection and author of the evolutionary theory of aging and senescence.

homeostasis and small body size required the early mammals to eat large amounts of food on a frequent basis, like living shrews. This requirement for a very high food intake may have caused the early mammals, like the living shrews, to live solitary lives so as to avoid intense competition for insects. Their small body size precluded the storage of energy reserves in the form of fat. Like living shrews, tenrecs, and opossums, they probably had large litters. Thus they may have faced much more intense intergenerational competition for limited food resources than had their cynodont ancestors, which may have led to the evolution of senescence as a mechanism to insure the survival of their offspring as postulated by Wallace.

Is rapid senescence the inevitable consequence of a high metabolic rate? The living shrews, tenrecs, and opossums have short life spans of only a year or two, but small primates like the mouse lemurs and the pygmy marmosets can live for 15 years. Moreover, birds, which have higher metabolic rates than do mammals, tend to live much longer than do mammals of similar body size. It may be significant that primates are arboreal and birds can fly. Both arboreality and flight greatly expand the foraging options for these animals, and as Steven Austad has observed, both are associated with longer life spans than in terrestrial animals. These observations indicate that the rates at which animals age are specific to particular groups of animals and may be related to their particular ecological circumstances.

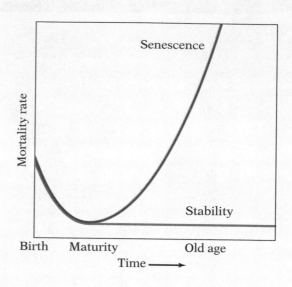

Senescence is reflected in the increased risk of dying with age, shown by the red curve. In 1825, the British actuary Benjamin Gompertz first described mathematically this increasing risk, the basis for the life insurance industry and all old age pension plans. Not all species have an increased mortality risk after maturity. Species lacking senescence follow the blue curve; their risk of dying does not increase with age. Some authors have mistakenly assumed that senescence does not occur in nature and that under natural conditions animals die from predation and disease shortly after maturity. Contrary to this assumption, there are many well-documented instances of old mammals in carefully observed natural populations. The rate of senescence varies greatly among mammals. Rapid senescence has been found in natural populations of opossums by Steven Austad; very slow senescence has been found in natural populations of capuchin monkeys by John Robinson. Opossums and capuchins are about the same body size; opossums are old at 2 years, but capuchins are barely out of infancy at 2 and are not old until 35 to 40. A capuchin lived to be 54 under human care; he is depicted on page 169.

THE HOMEOSTASIS NETWORK

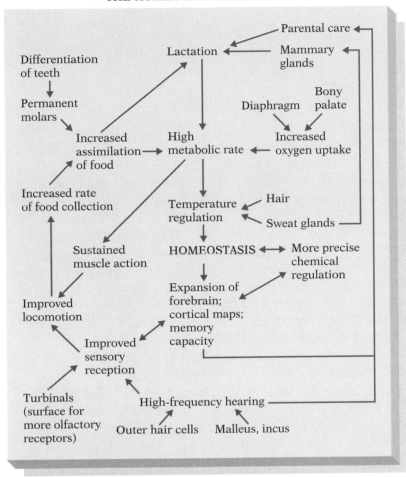

was slow continual growth in the adult phase as evidenced by the continual replacement of teeth throughout life.

The evolution of the capacity to maintain a constant body temperature in the mammalian line was the result of many interdependent adaptations. T. S. Kemp has observed that the changes must have occurred in tandem and that only small changes in any one system could occur without changes in related systems to support it. It is precisely this interdependence of adaptations that makes the study of evolution such a difficult intellectual challenge. One cannot isolate any single factor and declare it to be the "cause" responsible for the evolution of temperature homeostasis or any other adaptive complex. However, it is evident that the changes in the brain, and particularly those in the forebrain, were a crucial part of the set of adaptations necessary to maintain a constant body temperature in mammals.

## The First Birds

The other group of small predators to survive the great extinctions at the end of the Permian period were the thecodonts, which gave rise to the dinosaurs and birds. Unlike the rich fossil record for the cynodont antecedent to the first mammals, the fossil relics of the immediate ancestors of birds are quite scarce. One candidate for the earliest bird, Protoavis, lived in the late Triassic period, at about the same time as the first mammals. Protoavis weighed about 600 grams and was about the size of a pheasant. In contrast to the early mammals, with their reduced vision and enhanced sense of smell, Protoavis had very large eyes and a relatively poorly developed olfactory system, just the opposite condition. Also unlike the early mammals, it was probably active during the day. The brain of Protoavis was substantially enlarged relative to that of reptiles and was well within the range for living birds of its body mass. The brain enlargement was not necessarily associated with the capacity to fly. The size of the brains of ancient flying reptiles, Pterodactylus and Pteranodon, were within the range for living reptiles, while Troödon, a very advanced predatory dinosaur from the late Cretaceous, had a brain within the range for modern birds.

In Protoavis, the forebrain was well developed and possessed a well-defined anterior bulge called the wulst, which in living birds contains a topographic map of the visual field. The optic tectum also was highly developed in Protoavis. The famous fossil bird Archaeopteryx, which lived in the middle of the Jurassic about 150 million years ago, similarly had a brain size that is well within the range for living birds. We know that Archaeopteryx had feathers because their impressions appear in the fine-grained limestone in which the fossils were deposited. The presence of feathers for insulation in Archaeopteryx and the fact that all living birds are warm-blooded suggest that the ancient birds were temperature homeostats. Like the early mammals, the first birds would have needed a system of parental care to sustain growing infants. In contrast to mammals, where the mother caries the full energetic burden for sustaining her infants

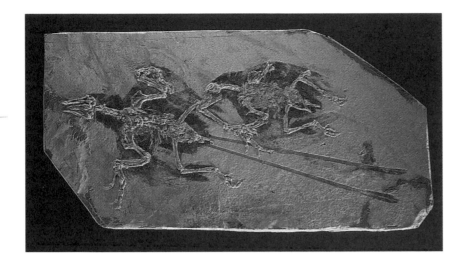

A pair of early birds, Confuciusornis sanctus, that lived 120 million years ago. The impressions of feathers surround the presumed male (left) and the female (the identification is based on the male bird's long tail feathers and larger size). Recently, a number of well-preserved fossils have been uncovered that are intermediates between birds and dinosaurs, although they are more recent in geological age than the earliest birds and thus are not ancestral to birds. However, these fossils do demonstrate a close affinity between birds and dinosaurs.

through lactation, in most birds both parents share in the provisioning of the growing offspring. Thus, biparental care was probably established in the earliest birds. The well-documented presence of insulating feathers in late Jurassic dinosaurs suggests that they too may have been temperature homeostats, and there is evidence from the work of Jack Horner and others that some of the dinosaurs provided parental care for their offspring.

## Uniquely Mammalian: The Neocortex

The first true mammal for which there is a brain endocast, Triconodon had a brain of a size that lies in the lower part of the range for modern mammals. There are many modern mammals, such as the shrews, tenrecs, and opossums, with relative brain sizes that are no larger than that of this ancient mammal. For this reason, these living mammals are sometimes said to occupy Mesozoic niches surviving from the times when the dinosaurs dominated the earth. In the transition from cynodonts to mammals, the relative size of the forebrain expanded. The neocortex, the sheetlike, six-layered structure in the roof of the forebrain that is found in all mammals and only in mammals, was probably present in the earliest true mammals; it is possible that it may actually have evolved earlier, at

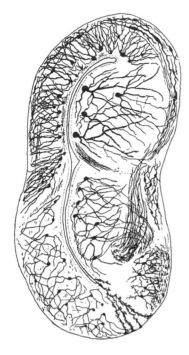

A section through the left forebrain of a frog stained with the Golgi method to reveal the cell bodies and dendrites of neurons; from the work of Pedro Ramón y Cajal. Note that the cell bodies of most of the neurons are located near the interior ventricle, the narrow space within the brain. The dendrites radiate out toward the exterior surface of the brain. This is the basic morphology of pyramidal neurons.

some point after the separation of the line leading to the mammals from the lines leading to reptiles and birds. The antecedents of the neocortex are present in the telencephalic roof in even the most primitive vertebrates. The neocortex is a specialization in the telencephalon that parallels the formation of the dorsal ventricular ridge and wulst in reptiles and birds. The neocortex is just as much a unique defining feature of mammals as are the mammary glands or the malleus and incus in the middle ear. As with the other distinctive features of mammals, the neocortex probably evolved as part of a set of adaptations related to temperature homeostasis. The large increases in metabolic expenditure necessary to sustain temperature homeostasis required commensurate increases in the acquisition of food by the early mammals. Since these animals were small and had only a limited capacity to store energy as fat, they were constantly under the threat of starvation. The neocortex stores information about the structure of the environment so that the mammal can readily find food and other resources necessary for its survival.

A structure probably resembling the antecedent to the neocortex can be seen in the dorsolateral telencephalon in amphibians. This is a sheet bounded by the ventricle on the inside and by the external surface of the brain. Located near the ventricle is a layer of pyramidal neurons with apical dendrites extending toward the outside surface of the brain. These dendrites receive input from the olfactory bulb and from the anterior thalamus, which relays visual, auditory, and somatosensory information from the tectum. Thus the dorsolateral telencephalon is a site of convergence of sensory input from the various modalities. These inputs probably form an asso-

The structure of the reptilian cortex, based on the work of Philip Ulinski. Pyramidal neurons are represented by the small triangles. In one, the dendritic arborization extending toward the cortical surface is illustrated. Thalamic fibers synapse on the outer parts of the pyramidal cell's dendritic tree. Fibers from other parts of the cortex synapse on the inner parts of the dendritic tree.

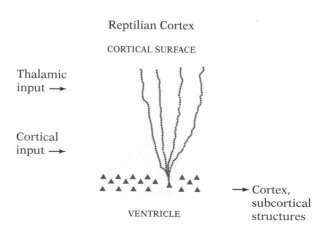

Reptilian Cortex

CORTICAL SURFACE

Thalamic input →

Cortical input →

→ Cortex, subcortical structures

VENTRICLE

ciative network that enables the amphibian to have some behavioral plasticity and adaptive response to stimuli.

The neocortex has the same location within the mammalian telencephalon, but it has five layers of pyramidal neurons with apical dendrites projecting toward the outside surface and a sixth outermost layer containing the tops of the apical dendrites. Each of the layers has a distinct set of connections with other parts of the brain. For example, the main input comes into layer 4; layer 6 sends feedback to

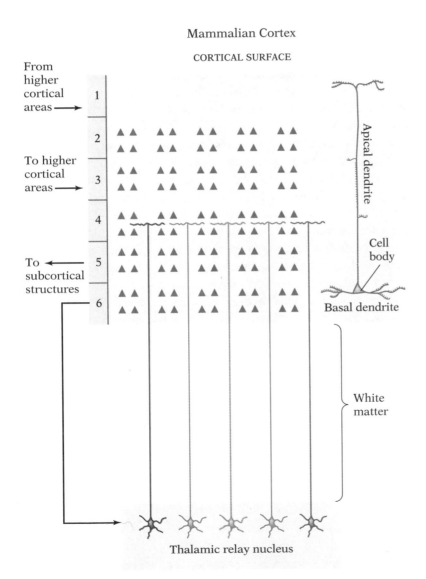

The structure of the mammalian neocortex. Compared with the reptilian cortex, there are additional layers of pyramidal neurons, and the inputs and outputs are layer specific. The input from the thalamic relay nucleus is topographically mapped as indicated by the color-coded input. An example of a thalamic relay nucleus is the lateral geniculate nucleus, which receives a topographic input from the retina and relays it to the primary visual cortex. The apical dendrites of pyramidal neurons often span much of the cortical thickness and tend to be oriented perpendicular to the cortical surface.

the source of input; output to other parts of cortex emerges from layers 2 and 3, and output to subcortical structures comes from layer 5.

Why is the neocortex a sheet rather than a globular structure? The sheetlike architecture of the neocortex may be imposed by the length of the apical dendrites of the pyramidal neurons that span the cortical thickness. The apical dendrites receive stratified synaptic inputs from different sources. The apical dendrites integrate these diverse inputs and are also influenced by action potentials relayed from the cell body. The biophysical mechanisms responsible for this integration may be able to operate over only a few millimeters, and this constraint may limit the thickness of the cortex.

Another major feature of neocortical organization is that the input to layer 4 from the thalamus is topographically organized. For example, the lateral geniculate nucleus, a structure in the thalamus, receives a topographic input from the retina and sends a topographic array of fibers that terminate in the primary visual cortex. Similarly, the somatosensory and auditory nuclei of the thalamus are linked topographically to the somatosensory and auditory areas of the neocortex. These topographic projections from the thalamus onto layer 4 of the cortex are responsible for the sensory maps. Most of the neocortex in the platypus, opossum, and hedgehog is devoted to topographically organized sensory maps. These findings suggest that neocortical maps are at least as old as the common ancestor of monotreme, marsupial, and eutherian mammals. In sum, these data suggest that topographically organized maps are an ancient feature of neocortical organization.

The neocortical neurons originate from a zone of dividing cells lining the ventricles of the fetal brain and migrate from the ventricular zone to the outside cortex along special guides known as the radial glia. The neocortex forms from the inside out, with cells in the deepest layers arriving in place before those of the more superficial layers. One of the mysteries concerning the neocortex is how the cortical areas are specified in embryogenesis. For example, there is considerable controversy as to the degree to which the spatial pattern of cortical precursor cells in the ventricular zone may determine the topographic organization of the neocortex. Pasko Rakic and others believe that a "protomap" within the ventricular zone heavily influences the topographic organization of the neocortex, while Otto Creutzfeldt and others believe that the neocortex is essentially a *tabula rasa* and that its topography is largely determined by the pattern of inputs from the thalamus. Work by Dennis O'Leary and his colleagues indicates that there is considerable plasticity in the formation of cortical areas, yet their basic topographic pattern is relatively constant among individuals of the same species and even among different species, suggesting a significant role for the genetic regulation of cortical development. Part of the specificity must come from the spatial ordering of the connections between the thalamus and the cortex. Some of the specificity may also arise from address signals within the cortex itself. Michel Cohen-Tannoudji and his colleagues have recently found a gene that may specify the development of the somatosensory cortex. This gene is part of the system that specifies cellular identity throughout the body, a family of genes called the major histocompatibility complex, which is expressed on the surfaces of cells. This gene is expressed heavily in layer 4 of the developing somatosensory cortex and in only a very sparse scattering of cells elsewhere in the brain.

Cohen-Tannoudji and his colleagues removed the cortex containing the site of gene expression in embryos before the cortex received input from the thalamus onto layer 4. They transplanted these pieces of cortex into other parts of the brain and found that the transplanted cortex expressed the gene. Thus the gene was not dependent on the local environment for its expression before birth, and it is a good candidate for the agent responsible for specifying somatosensory cortex in embryogenesis. It will be interesting to determine whether other cortical areas also have cell-surface molecules that can be linked to their development, and whether they specify the identity of cortical areas in embryogenesis.

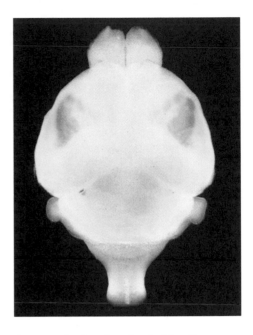

A mouse brain, showing in blue the pattern of expression of the gene *H-2Z1*. The location of the gene expression corresponds to somatosensory cortex. Yorick Gitton, Michel Cohen-Tannoudji, and Marion Wessef discovered that this gene is expressed in somatosensory cortex tissue transplanted to other parts of the brain before birth, but that after birth the expression of this gene depends on the presence of thalamic fiber input.

## The Mouth Leads the Way in Cortical Development

In embryos the different parts of the neocortex do not develop simultaneously. The first region of the neocortex to develop in the fetus is the part that will become the representation of the mouth and tongue in the somatosensory and motor cortex. The neocortex then develops in concentric zones extending out from this core region. The early development of the mouth representation in the somatosensory and motor cortex is probably related to the need by young

Sucking the thumb may lead to the increased development of the cortical representation for that hand and perhaps to the dominance of the hemisphere containing that representation.

mammals to nurse as soon as they are born. Thus the mouth representation may be the seed crystal around which the other parts of the neocortex form during embryogenesis. Ultrasonic images of the womb often show primate fetuses sucking on a thumb. This sucking action may stimulate the formation of the cortical maps of the mouth and hand in the primate fetus as connections form late in fetal development. If the fetus consistently sucked on the thumb of one hand as opposed to the other, the increased stimulation might favor the development of its cortical representation, which in turn might lead to hand preference and perhaps even to a greater development of the hemisphere containing the more developed representation. Thus the preferential sucking of one thumb might lead the cerebral dominance. This theory could be easily tested by observing fetal thumb-sucking with ultrasound and determining whether it predicts hand preference later in life.

# Diverging Patterns in the Telencephalon of Mammals, Reptiles, and Birds

The organization of the telencephalon took an entirely different course in reptiles and birds than it did in mammals. A recent embryological study by Anibal Smith Fernandez and his colleagues has done much to clarify the diverging patterns of telencephalic development. They studied cell migration and compared the patterns of expression of marker genes at different stages in the embryonic development of frogs, turtles, chicks, and mice. The genes are ones we have encountered before, *Emx-1* and *Dlx-1*. In all four vertebrates, the overall pattern of expression of these genes in the telencephalon is much the same. *Emx-1* is expressed dorsally; *Dlx-1* is expressed ventrally, and there is a smaller intermediate zone between them.

The dorsal region (red in the illustration below) in mammals becomes the cerebral cortex (cx), including both the neocortex and

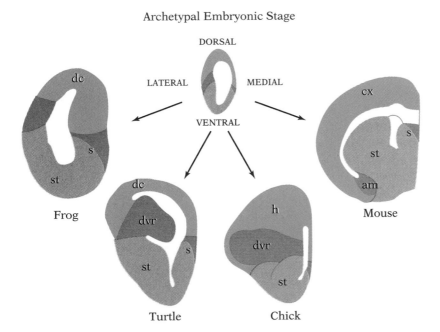

Archetypal Embryonic Stage

The evolution of the telencephalon based on the expression patterns of regulatory genes during embryonic development, from the work of Anibal Smith Fernandez and his colleagues. In the frog, the dorsal telencephalon is also termed the pallium, from the Latin work for "cloak."

hippocampus; in reptiles it becomes the dorsal cortex (dc) and hippocampus; in birds it becomes a structure known as the hyperstriatum (h), which includes the wulst.

The intermediate region (blue in the illustration on page 113) on the lateral side in mammals becomes the laterobasal amygdala (am), which is involved in emotional learning and in humans has an important role in the perception of emotion in facial and vocal expression. In reptiles and birds, the intermediate zone bulges into the ventricular space and is known as the dorsal ventricular ridge (dvr). Harvey Karten has proposed that the dorsal ventricular ridge is homologous with a portion of the sensory cortex in mammals because these structures share similar input from another part of the brain located in the thalamus. However, as pointed out by Laura Bruce and Timothy Neary, there are also similar connections between the thalamus and the laterobasal amygdala that are consistent with the gene expression patterns, indicating a homology between the laterobasal amygdala in mammals and the dorsal ventricular ridge. The intermediate zone on the medial side of the embryonic telencephalon in all these animals becomes part of the septum (s), which is a structure closely linked to the hippocampus.

The ventral zone (green in the illustration on page 113) in all these animals becomes the striatum (st), which has an important role in the control of the muscles. The common patterns of gene expression in each of these zones implies that the structures within each are homologous in different vertebrates, that is to say, they were derived from the same structure in their common ancestor.

The dorsal cortex in turtles retains the ancient pyramidal neuron architecture seen in amphibians. Part of this cortical structure receives an input from the lateral geniculate nucleus, which in turn receives its input from the retina. However, the visual cortex in reptiles contains only a very crude topographic map of the retina. In birds, there is a considerable expansion of this part of the telencephalon, especially in the wulst. Harvey Karten and his colleagues and Jack Pettigrew found that the wulst contains a single, highly topographic field map in owls. Pettigrew also observed that the wulst contains no pyramidal neurons but instead stellate, or star-shaped, cells that lack a large apical dendrite extending toward the surface of the brain; the dendrites extend in all directions. The stellate cell architecture is a specialization of the wulst in birds. The basic cortical pyramidal cell architecture is more primitive since it is present in the homologous telencephalic structures in amphibians and reptiles.

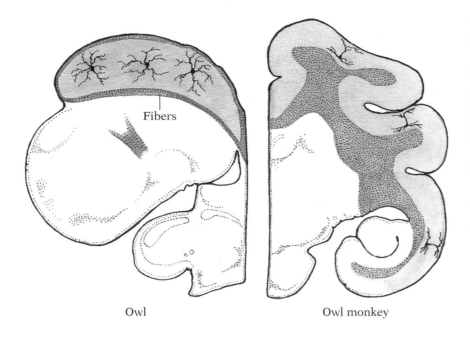

Fibers

Owl

Owl monkey

A comparison between the wulst in the owl and the neocortex in the owl monkey. The neocortex is a folded sheet, which in cross section appears as a ribbon. Underlying the neocortex is an extensive region of white matter containing the fibers connecting the different parts of the neocortex. The neocortical white matter constitutes a large fraction of the forebrain volume. Longer connections linking cortical areas not only take up more space but also require more time for the transmission of information between areas. The wulst is much thicker than the neocortex and appears to be wired much more efficiently.

These observations imply that there was a fundamental change in the wiring of this region of the brain in the evolutionary line from reptiles to birds. The architecture of the wulst is not constrained by apical dendrites, and this is perhaps why the wulst expands to a much greater thickness than the neocortex does in mammals.

Thus in both birds and mammals maps formed from part of the dorsolateral telencephalon; however, the principal neuronal constituents and architecture of the neocortex and wulst are quite different. In the neocortex there are multiple visual maps laid out in a relatively thin sheet, whereas in the wulst there is a single map in a much thicker structure. One consequence of multiple maps in mammals is that the connections between the maps must traverse relatively long distances and thus in larger brains a greater amount of space must be devoted to the neocortical white matter that carries the "wires" connecting the maps. The avian architecture is far more economical in terms of the wiring because it involves only a single large map in the wulst.

## The Wiring Cost of Expanding Neocortex

One cost of expanding neocortex can be measured by comparing the neocortical gray matter, which contains the active elements in neural computations, with the underlying neocortical white matter, which contains the "wire," the axon fibers that connect the different parts of the neocortex. The volume of neocortical white matter increases at the 1.318 power of the volume of neocortical gray matter. This is very close to a 4/3 power law relationship, which suggests that there may be simple geometrical factors that govern the increasing size of the white matter.

Thus as the neocortex expands an increasingly larger part of the brain must be devoted to the wires connecting it. The neocortical white matter is like the infrastructure that supports our economy. Like our telecommunications systems, the neocortical white matter does not make the decisions that run the system, but it is necessary for its functioning. With systems of increasing complexity, an increasingly larger part of the total system must be devoted to infrastructure.

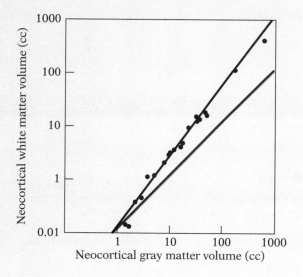

The expansion of neocortical white matter relative to neocortical gray matter. As the size of the neocortical gray matter increases, the size of the associated white matter grows disproportionally; a proportional relationship would follow the red line. The volumetric data were obtained from the work of Heinz Stephan and his colleagues. The analysis was performed by Andrea Hasenstaub and the author.

The great increase in energy metabolism puts enormous demands on the cognitive and memory capacity of the brain in warm-blooded vertebrates because they must locate large amounts of food on a regular basis. The formation of highly ordered sensory maps in the dorsolateral telencephalon in mammals and birds is part of the adaptive complex supporting the greater energy requirement to sustain temperature homeostasis.

All young mammals play, which is probably crucial for the maturation of the cortex and the learning of adult roles. Here young wolves engage in social play.

## Forebrain Expansion and Memory

Young reptiles function as miniature versions of adults, but baby mammals and birds are dependent because of their poor capacity to thermoregulate, the consequence of their need to devote most of their energy to growth. Most mammals solve the problem with maternal care, shelter, warmth, and milk. In most birds, both parents cooperate to provide food and shelter to their young. The expanded forebrain and parental care provide mechanisms for the extra-genetic transmission of information from one generation to the next. This transmission results from the close contact with parents during infancy, which provides the young with opportunity to observe and learn from their behavior; the expanded forebrain provides an enhanced capacity to store these memories. The expanded forebrain

A juvenile Japanese macaque monkey engages in object-oriented play.

and the observation of parents are probably necessary for the establishment of successful caregiving behavior itself, as the young mature into adults that will in their turn have to serve dependent young. During the period of infant dependency, baby mammals and birds play, behavior that may be essential for the development of the forebrain. The baby's playful interaction with its environment may serve to provide the initial training of the forebrain networks that ultimately will enable the animal to localize, identify, and capture resources in its environment. In humans this playful interaction persists into adulthood, in perhaps another example of pedomorphy in our evolutionary history.

Although synapse formation and modification in the forebrain occur throughout life, these processes are most active during the period in which the infant is dependent on its parents for sustenance. Yoshihiro Yoshihara and his colleagues have recently discovered a possible key molecular participant in these processes, which they have termed telencephalin. This protein is found only in the telencephalon, and it begins to form just before birth. Telencephalin is related to the class of proteins known as cell adhesion molecules, which serve to establish connections between cells. It spans cell membranes and extends in a series of loops into the space just outside the cell. Telencephalin is located in the membranes of dendrites and cell bodies but not axons and thus corresponds to the sites where neurons receive synaptic contacts.

The exact means for how telencephalin participates in synapse formation and modification has not been established, but very recently an analogous system has been discovered in fruit flies that suggests a possible mechanism. Fruit flies learn to avoid odors that are paired with electric shocks, and this learning depends on specific structures in their brains called the mushroom bodies. Michael Gotewiel and his colleagues have discovered in flies a mutant gene called *Volado*, a name derived from a Chilean slang expression for being forgetful. This mutant reduces the fly's ability to learn to avoid shock-paired odors, and it resembles the learning deficit produced by damaging the mushroom bodies. *Volado* encodes a protein that is a member of the cell-adhesion family and is located in the synaptic regions of the mushroom bodies. In a very ingenious experiment, Gotewiel and colleagues showed that *Volado* did not injure the mushroom bodies. They introduced into *Volado* mutants the nonmutated (wild-type) form of the *Volado* gene in such a way that it could be turned on for a few hours, after which it turned off. During the period in which the nonmutated gene was activated, the fly learned normally; afterward it was as forgetful as before. Their results suggest that learning in flies depends on the proper functioning of a cell-adhesion molecule located in the synaptic membrane. The existence of a similar molecule associated with synapses in the telencephalon implies that this mechanism may also operate in that structure as well. Thus memory may depend on the physical tightening or loosening of the synaptic connections between neurons.

Telencephalon

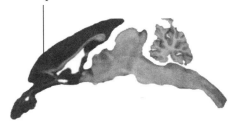

The dense black staining indicates the distribution of telencephalin in the brain of a mouse; from the work of Y. Yoshihara and his colleagues. The distribution of telencephalin, stained dense black, corresponds with the telencephalon in the mouse brain.

## Warm-Blooded Paradoxes

There are several paradoxes in the evolution of warm-blooded vertebrates. One is that the evolution of temperature homeostasis in the early mammals was probably associated with shorter individual life spans than those enjoyed by their cynodont forebears. The short life spans of the early mammals was probably related to the very high energy costs of temperature homeostasis in a small animal. Small mammals must eat a great deal to support homeostasis and generally can go only a short time without eating. Therefore the early mammals were at great risk of starvation, which was the price of being able to function independently of variations in environmental temperature. Thus the early evolution of mammals was a trade-off between two buffers against environmental variation: there was an expansion of the buffer against temperature variation at the expense of increased vulnerability to fluctuations in food resources over time. Birds managed to escape this painful trade-off. Birds live much longer than mammals of the same size, possibly because the early mammals had much more restricted foraging opportunities than the birds, which were capable of flight. This would have enabled birds to exploit a much wider variety of resources to sustain their higher energy requirements. Temperature homeostasis potentially opened up many new niches for animals that were less constrained by variations in environmental temperature. With these remarkable innovations in mammals and birds, one might think that they would have been immediate successes. This is far from what actually happened, and this is another paradox in the evolution of temperature homeostats. Both mammals and birds enjoyed only modest success for many millions of years. It was not until another wave of mass extinctions occurred in the global winter at the end of the Cretaceous period, 65 million years ago, in which the dinosaurs and many other animals disappeared, that mammals and birds finally began to develop the enormous diversity that we see today.

Watercolor of a young orangutan by Richard Owen, a sensitive artist as well as a great comparative anatomist.

# Primate Brains

In the distant future I see open fields for far more important researches. Psychology will be based on a new foundation, that of the necessary acquirement of each mental power and capacity by gradation. Light will be thrown on the origin of man and his history.

Charles Darwin,
*The Origin of Species*, 1859

The 6-mile-wide meteorite that struck Yucatán 65 million years ago caused the earth to be enveloped in a huge cloud of dust and debris that blocked sunlight for many months. This event destroyed the dinosaurs and many other groups of animals. The mammals, however, were well equipped to survive this cold, dark period because they were active at twilight or at night, they were warm-blooded, and they were insulated with fur. When the dust finally settled the mammals found a world in which most vertebrates larger than themselves were dead: the meek had inherited the earth. From the stock of early mammals new forms emerged to seize the niches vacated by the lost animals. Other mammals, including our ancestors, the early primates, created new niches for themselves in this fundamentally altered environment. Once the dust settled, that environment became much warmer than today's world, and tropical rain forests covered a much larger portion of the planet than they do now.

## Eyes, Hands, and Brains

The early primates lived in these forests and started to become abundant about 55 million years ago. Much is known about these early primates because they left behind many fossils: they are closely related to the group of living primates called the prosimians, a name that means "before the monkeys." The prosimians include the tarsiers, galagos, lorises, and lemurs. The early primates weighed only a few ounces, and they clung with their tiny grasping hands to the fine terminal branches of trees in the tropical rain forest. Their large eyes faced forward, and their visual resolving power was greatly improved by an increased density of photoreceptors in the center of their retinas. Emerging from this dense array of photoreceptors was a strong set of connections from the central retina via the optic nerve to the brain. The visually mapped structures in the brain contained greatly expanded representations of the central retina. In some of these structures there was a marked segregation of visual processing into two distinct functional streams, one exquisitely sensitive to motion and small differences in contrast, the other responsive to the shape and form of visual objects. The visual cortex, the major site of visual processing in the brains of primates, enlarged greatly, and many new cortical visual areas formed that were not present in the primitive mammals. Another innovation in the early primates was a specialized cortical area devoted to the visual guidance of muscle

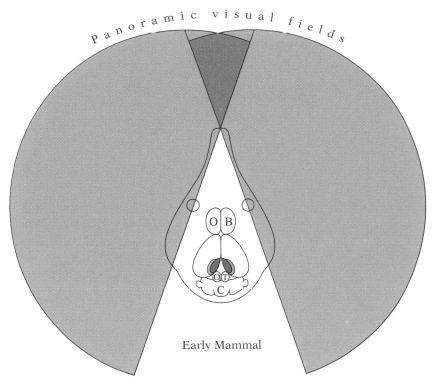

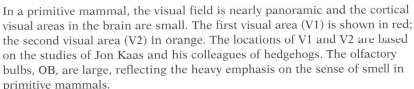

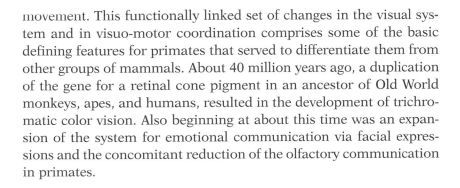

Early Mammal

A hedgehog, *Erinaceous europaeus,* a living nocturnal insectivore that has retained many features of primitive mammals.

In a primitive mammal, the visual field is nearly panoramic and the cortical visual areas in the brain are small. The first visual area (V1) is shown in red; the second visual area (V2) in orange. The locations of V1 and V2 are based on the studies of Jon Kaas and his colleagues of hedgehogs. The olfactory bulbs, OB, are large, reflecting the heavy emphasis on the sense of smell in primitive mammals.

movement. This functionally linked set of changes in the visual system and in visuo-motor coordination comprises some of the basic defining features for primates that served to differentiate them from other groups of mammals. About 40 million years ago, a duplication of the gene for a retinal cone pigment in an ancestor of Old World monkeys, apes, and humans, resulted in the development of trichromatic color vision. Also beginning at about this time was an expansion of the system for emotional communication via facial expressions and the concomitant reduction of the olfactory communication in primates.

A prosimian, the slender loris, *Loris gracilis*, using its prehensile hands and feet to cling to a fine branch. By occupying the fine-branch niche primates gained access to a rich array of resources such as fruit and insects, but living in this precarious environment requires superb vision and visuo-motor coordination.

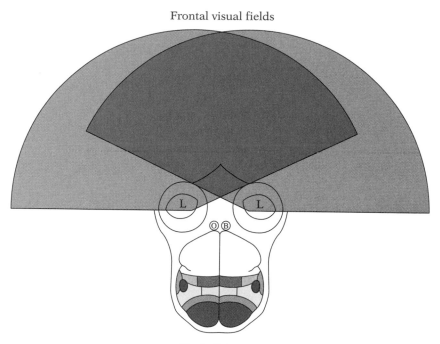

Frontal visual fields

Early Primate

In a primitive primate, the large eyes are directed forward and there is a large amount of binocular overlap between the visual fields of the two eyes (L = lens). The olfactory bulbs are smaller than in primitive mammals. The first visual area (V1) is shown in red; the second visual area (V2) in orange; the third tier of visual areas in yellow; area MT in dark blue; the inferotemporal visual cortex in green; the posterior parietal cortex visual cortex in brown; the temporoparietal visual cortex in purple. The positions of the eyes and the locations of the visual areas are based on the author's studies of prosimians with high-resolution magnetic resonance imaging and neurophysiological recording, and on the remarkably well preserved skulls and brain endocasts of Eocene primates.

## The Advantages and Costs of Front-Facing Eyes

Front-facing eyes and the expansion of the size and number of cortical visual areas are distinctive features of primates and are related to the primate capacities for keen vision and eye–hand coordination. Two theories have been proposed to explain the development of high-acuity frontal vision and eye-hand co-ordination in primates: Matthew Cartmill's "visual predator" hypothesis and Robert Martin's "fine-branch niche" hypothesis. The two theories are not mutually exclusive. Cartmill has suggested that the early primates were hunters who relied mainly on vision. He based this inference on the fact that many small prosimian primates, such as tarsiers and mouse lemurs, capture and eat insects and small vertebrates, and that nonprimates with large front-facing eyes, such as cats and owls, are predators. Martin has proposed that the early primates used their grasping hands to move about in the fine branches of the forest canopy and exploit the rich abundance of fruit and insect resources available there. Keen vision and superb eye–hand coordination are required to function in the fine-branch niche, a uniquely complex visual environment in which the branches move and sway, where the penalty for miscalculation can be a fatal fall.

What advantages do front-facing eyes provide to predators? Because of the bilateral symmetry of their limbs, predators generally orient themselves so that their prey is located directly in front of them and they can propel themselves swiftly forward, carrying out a coordinated attack with forelimbs and jaws. Frontally directed eyes provide maximal quality of the retinal image for the central part of the visual field. This is where the prey is located in the crucial moments, just before the final strike, when the predator is evaluating the prey's suitability as food, its evasive movements, and its ability to defend itself. Image distortion tends to increase the farther an object is located off the optical axis of a lens system, and thus it is advantageous to a visually directed predator to have front-facing eyes in which the optical axes are directed toward the central part of the visual field. Such image distortion can be reduced by decreasing the aperture of the lens, but the early primates were probably active at twilight or night when light was at a premium, and the larger aperture was needed to collect as much light as possible. Indeed, the familiar examples of nonprimates with large front-facing eyes are cats and owls, both night hunters.

Front-facing eyes also increase the size of the binocular visual field, enhancing visual capabilities in at least three ways. The first is by the expansion of the stereoscopic visual field. Objects cast slightly different images in each of the two eyes. The visual cortex is sensitive to these small differences, which it interprets as relative differences, in depth. Stereoscopic depth perception provides a relative measure of distance that can guide a predator in seizing its prey.

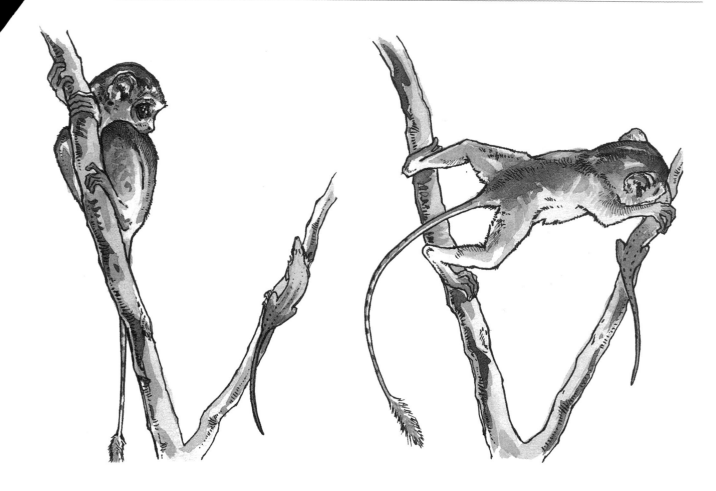

A tarsier preying upon a lizard. These drawings are based on photographs taken by Johannes Tigges and W. B. Spatz.

Another important function of binocular vision, pointed out by Bela Julesz, is to "break" camouflage. Prey often adopt the protective strategy of matching themselves to their background and are difficult to detect monocularly; the binocular correlation of the images from both eyes may enable the predator to detect prey thus concealed. Finally, under low light conditions, the binocular summing of images from both eyes can facilitate the detection of barely visible prey.

Moving about in the fine terminal branches also requires keen vision in the part of the visual field immediately in front of the ani-

mal. Thus, the fine-branch-niche hypothesis for the origin of primates shares with the visual-predator hypothesis the necessity for high-quality vision in the space immediately in front of the animal, where it can manipulate objects with its hands. Thus both hypotheses predict front-facing eyes for improved image quality and stereoscopic depth perception. However, living in the trees does not in itself lead to front-facing eyes. Squirrels are highly adept at moving from branch to branch in the trees, yet they have laterally directed eyes with nearly panoramic vision. Still, the squirrel's small strip of binocular visual field has a large representation in the visual cortex, suggesting that binocular vision and perhaps stereopsis may be important to the squirrel even though its eyes are laterally oriented.

Along with the advantages they confer, front-facing eyes impose a significant cost on primates because the nearly panoramic visual field found in most mammals is constricted, and the ability of primates to detect predators approaching from the rear is limited. This constriction of the visual field predisposes primates to develop other means for detecting predators. Some prosimians, such as the galagos, which can direct the orientation of their ears with delicate precision, have the keen ability to detect the sources of sounds that might signal a predator's approach. The early primates, like most primitive mammals, probably lived a solitary existence. However, the loss of panoramic vision strongly favored the formation of social groups because multiple sets of eyes could overcome the vulnerability imposed by the restriction of the visual field. The response to this limitation may have been the evolution of neural systems for social cooperation and the production of vocalizations that signal the presence of predators. Dorothy Cheney, Robert Seyfarth, and others have found that primates have specific alarm cries for aerial versus ground predators. The evolution of these specific alarm cries presents something of a puzzle since the animal making the cry calls attention to itself, which might increase its risk of being attacked by the predator. It has been suggested that such apparently altruistic acts, while possibly endangering the crier, increase the chances of survival of close relatives that share most of the genes possessed by the animal making the alarm cry. Thus the cooperative behavior enhances the chances that those shared genes will be passed on to the next generation.

## The Optic Tectum: An Ancient Visual System Transformed

The midbrain in primates contains the ancient visual map, the optic tectum, found in all vertebrates. In nonprimates, the optic tectum on one side receives most of its fibers from the retina on the opposite side, and the primitive condition is a nearly completely crossed projection from the retina to the optic tectum. In primates, the front-facing eyes have caused a remodeling of the connections between the retina and the optic tectum: there is a large projection from the

The mapping of the visual field on the optic tecta in primates and nonprimates. The star indicates the center of the visual field; the small circles indicate the vertical midline of the visual field, which separates it into right and left hemifields. The monocular crescent is the part of the visual field that is seen by only one eye. Each tectum—there is one on each side—is the dome- or disk-shaped structure forming the roof of the midbrain. In the diagram, the anterior edge is at the top of each tectum. In primates the representation of the vertical midline of the visual field corresponds to the anterior edge of each tectum: the right visual hemifield is represented in the left tectum, and vice versa. In nonprimates the representation in the anterior tectum extends well beyond the vertical midline, and this part of the visual field is represented redundantly in the optic tectum on both sides of the midbrain.

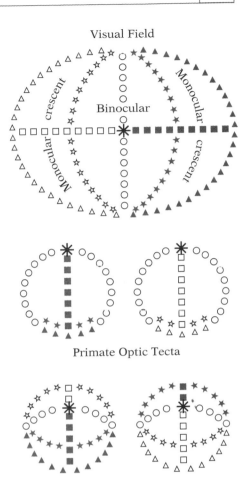

Visual Field

Primate Optic Tecta

Nonprimate Optic Tecta

retina to the optic tectum on the *same* side, and the maps in the tecta have been modified so that only the opposite half of the visual field is represented in each tectum. This change may have come about because the standard vertebrate tectal mapping would have resulted in redundant representations of the visual field in the tectum in primates. However, other animals with frontally facing eyes, cats and owls, have retained the same type of visual mapping that is found in other vertebrates rather than the modified version found in primates, and thus there is some redundancy in the tectal maps in these animals. In fish, amphibians, and reptiles the optic tectum has a broad array of functions consistent with its role as the main visual processing center in the brain. The tectum also serves to integrate visual, auditory, and somatosensory inputs. In primates the function of the optic tectum is more specialized, serving to guide the eyes so that the images of an object of interest fall directly on the central retina in the region of maximum acuity, which is only a very small part of the total retinal area. Thus a major function of the optic tectum in primates is to cause the eyes to fixate on interesting objects.

Primates fixate mainly by eye movements rather than head movements; by contrast, owls and cats rely mostly on head movements to look at interesting things. Part of the visual field map is represented in both sides of the tectum in nonprimates, but not in primates. Perhaps the nonredundant map found in the primate tectum reflects its role in directing the eye to fixate on objects of interest. Tectal map

Peter Schiller and Michael Stryker found a direct correspondence between the visual and visuo-motor maps in the optic tectum in monkeys. They mapped visual receptive fields (the circled areas) in the tectum and then electrically stimulated these sites. Stimulation caused the monkey to direct its eyes so that the center of the retina gazed at the site in the visual field corresponding to the previously recorded receptive field. (The eye movement is indicated by the arrows from the stars marking the fixation point to the receptive fields.) This visuo-motor response is a major function of the optic tectum in primates, causing the animal to look at novel objects that have entered its peripheral visual field. If primates had a redundant visual field map, as do other mammals, there would be ambiguity, indicated by the dashed circles and arrows, in the visuo-motor map that guides fixation. Redundancy in the visuo-motor map might compromise the primate's ability to fixate rapidly and accurately on novel objects.

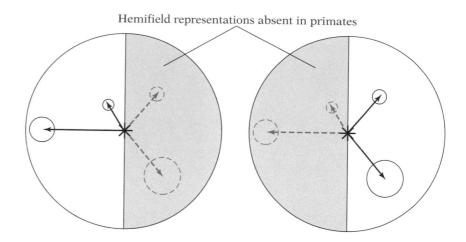

Hemifield representations absent in primates

redundancy in primates might have interfered with fixation by providing a superfluous target within the visuo-motor map.

The organization of the optic tectum is fundamentally transformed in primates, but is this transformation unique to primates? In 1977, I suggested that the organization of the optic tectum might be a defining feature that distinguishes primates from nonprimate mammals. A few years later Jack Pettigrew reported that *Pteropus*, a type of large bat from the group known as the megachiropterans, had the primate type of tectal map. He used this observation to argue that this group of bats were "flying primates." His proposal caused a considerable uproar among evolutionary biologists since it would have required a major revision of the basic system for classifying mammals by separating the megachiropterans from the smaller bats (microchiropterans) and lumping them with the primates. The heat of this controversy resulted in several scientific symposia and two detailed mapping studies of the optic tectum in the megachiropterans, *Rousettus* and *Pteropus*. Unlike primates, in which the visual field in each side of the tectum extends only to the midline, both studies found that the representation extended considerably beyond the midline into the visual field on the same side. These parts of the tectal map beyond the midline are thus represented on both sides and are redundant. Bats, like most mammals, have laterally placed eyes and low acuity. Comparative anatomical and DNA data also suggest that megachiropterans are more closely related to microchiropterans than to primates. Thus megachiropterans possess the nonprimate pattern of tectal organization and are unlikely to be "flying primates."

# Seeing Motion and Form

The analysis of images requires tracking and identifying objects. The great expansion of the visual system in primates occurred mainly in the forebrain, where two distinct systems evolved for seeing the motion and the form of objects in the visual scene. In primates the major output from the retina travels in the optic nerve to the lateral geniculate nucleus in the thalamus, which in turn connects with the visual cortex. "Nucleus" is the anatomical term for an aggregation

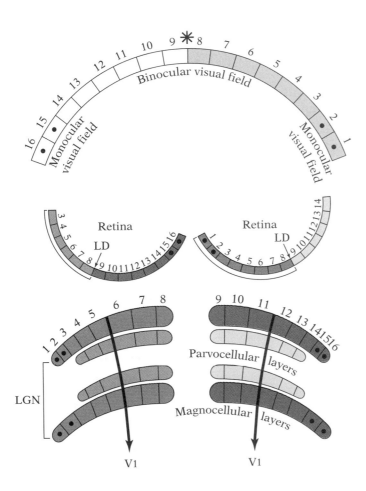

The mapping of the visual field onto the lateral geniculate nucleus in primates. The upper diagrams are horizontal slices through the visual field and the retinas. The retinas are divided into hemiretinas by the line of decussation (LD), or crossing, which corresponds to the vertical midline of the visual field. The green and blue hemiretinas, which view the right half of the visual field, project to the layers of the lateral geniculate nucleus on the left side of the brain. Note that the green input is not quite complete, because it does not include the monocular segment. Each hemiretina projects to an individual layer, and the layers are stacked in such a way that same places in the visual field fall in register. The parvocellular layers contain many small neurons, and their responses are specialized for fine detail in the visual image and in day-active primates for the analysis of color. The magnocellular layers contain fewer and larger neurons, and their responses are specialized for the analysis of low-contrast moving images. The axons of geniculate neurons project with a high degree of topographic precision onto layer 4 of the primary visual cortex (V1). The parvocellular and magnocellular cells project on different sublayers within layer 4, indicating a certain degree of parallel processing of the visual inputs, within the visual cortex.

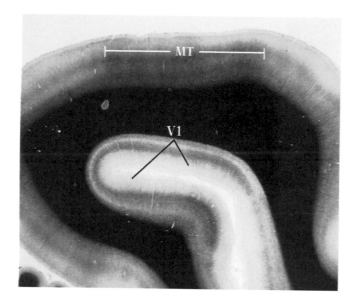

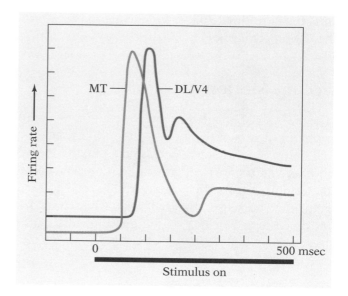

Left: A myelin-stained section through the brain of an owl monkey illustrates the distinct patterns in MT and V1. The myelinated rectangular fibers appear dark blue. The "white matter" stains black because it is made up almost entirely of myelinated fibers. Area MT is the dark blue rectangular band at the top of the section. The part of V1 that is buried in the calcarine fissure is located in the center of the section and shows the strong banding pattern that has led to its alternative name as "striate" cortex. Right: The responses of MT and DL/V4 neurons to a stimulus presented in their receptive fields. The stimulus was an optimally oriented bar. In each case the curve represents the summed responses for a population of neurons recorded from that area. Note that the responses recorded from MT rise and drop off much faster than do those recorded from DL/V4. The recordings were made in owl monkeys by Steven Petersen, Francis Miezin, and the author.

of neurons; "geniculate" derives from the Latin word for "knee" and refers to the shape of the nucleus; "lateral" refers to its location on the side edge of the thalamus. The lateral geniculate nucleus consists of several sets of layers, each of which receives fibers from either the eye on the same side or from the eye on the opposite side of the head. The layers are further specialized for function. One set, the magnocellular ("large-cell") layers, contains large neurons that receive input from the largest retinal ganglion cells with thick, fast-conducting axons. The magnocellular layers are sensitive to rapid movement and minimal contrast in light intensity. The second set, the parvocellular ("small-cell") layers, contains smaller neurons that receive input from smaller retinal ganglion cells with thinner, slower-conducting axons. The parvocellular layers detect finer detail but are less sensitive to motion and contrast than are the magnocellular layers. A partial segregation of the magnocellular and parvocellular inputs is maintained at higher levels in the visual pathway.

The magnocellular layers project to a separate layer of the primary visual cortex and thence via rapidly conducting axons to the middle temporal visual area, known as MT, where the neurons are very sensitive to the direction of visual motion. The perception of motion requires the fast conduction of the visual input. The fast conduction of information to MT is related to its function in the perception of movement. The speed of axonal conduction is related to the size of the axon and to the thickness of the myelin insulation: the magnocellular neurons have large axons, and the axons in area MT are thickly myelinated.

MT in turn projects to higher cortical areas in the posterior parietal lobe. The studies of Michael Goldberg, William Newsome, Richard Andersen, and their collaborators indicate that the posterior parietal lobe uses the visual input as part of a system to plan movements of the eyes and hands. A parallel stream of connections emerging from the primary visual area is made up of a more slowly conducting set of axons that relays a mixture of parvocellular and magnocellular inputs to the second visual area (V2) and thence to the fourth visual area (V4), where the neurons are very sensitive to size and shape of visual stimuli. Area V4 projects to the inferotemporal visual cortex, which is crucial for the visual memory of objects.

## Seeing Spots, Lines, and Curves

In 1958, David Hubel and Torsten Wiesel discovered that most neurons in the primary visual cortex are exquisitely sensitive to the orientation of straight lines and edges within their receptive fields. They also found that neurons within a vertical column extending from the cortical surface to the underlying white matter shared the same preferred line orientation. Neurons that specifically responded preferentially to particular orientations were soon found in other cortical visual areas that received input either directly or indirectly from the primary visual cortex. They also discovered neurons that

An analysis of Paolina, as described by Benoit Dubuc and Steven Zucker. The first image is a photograph of a statue of Paolina Bonaparte, Napoleon's sister, by Antonio Canova. The following four images show how visual cortical neurons could analyze the original image, based on the comparison within the receptive field between points on a line and the flanking regions around the line. Where there are no flanking regions, there are only points, or "dust." Where there are lines without flanking regions, there are curves. Where there are flanking lines at many different orientations, there is turbulence. Where there are flanking parallel lines, there is flow. Thus the system of oriented line detectors can analyze the underlying physical processes that create the visual scene.

Paolina

Dust

Curves

Turbulence

Flow

responded to the ends of lines and to corners defined by intersecting lines. These are probably the basis for the detection of the curvature of lines and the recognition of shapes. Their Nobel-prize–winning work is beautifully recounted in David Hubel's book *Eye, Brain, and Vision*. In 1980, Steven Petersen, Jim Baker, and I found that most neurons in V4 are very sensitive to the dimensions of the stimulus, with some neurons preferring tiny spots while others preferred long rectangles. This set of stimulus preferences by cortical neurons has intrigued theoreticians. Anthony Bell and Terrence Sejnowski have suggested that the orientation selectivity of cortical neurons is a computationally ideal system for analyzing the image properties of natural scenes. Benoit Dubuc and Steven Zucker have proposed that the detection of line endings and curvature form the basis for the visual analysis of complex objects.

## Area MT and the Perception of Motion

Area MT, which is present in all primates, is devoted to the analysis of movement in visual images and is one of the clearest examples of the specialization of function in the neocortex. MT also provides some of the best evidence that links neuronal activity to perception. In 1968, Jon Kaas and I first mapped the representation of the visual field in MT and found that it corresponds to a zone of the cortex that contains thickly myelinated axons. Shortly thereafter, Ronald Dubner and Semir Zeki found that the neurons in MT are very sensitive to the direction of movement of stimuli within their receptive fields. MT neurons respond maximally to a preferred direction and are often inhibited by movement in the opposite direction. Like the orientation-selective neurons in V1, the directionally selective neurons in MT are organized in vertical columns. Thomas Albright found that these columns are adjoined by columns containing cells with the *opposite* directional preference.

Adjacent columns with opposite preferred directions appear to be joined in such a way that activity in one suppresses activity in its antagonistic mate. This relationship is probably responsible for the striking motion aftereffect known as the waterfall illusion. This powerful illusion is elicited if you watch a waterfall for a minute or two and then direct your gaze to the nearby rocks, which will incredibly appear to move *upward* in the direction opposite to the falling water.

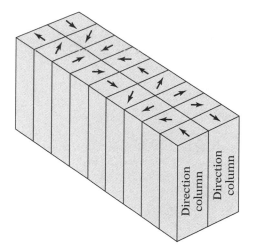

The columnar organization for direction preference in MT, based on recordings done by Thomas Albright. Each column extends from the cortical surface to the underlying white matter. Along one axis the directional preferences change gradually, but along the other axis adjacent columns are maximally responsive to opposite directions.

The illusion results when you have exhausted the MT neurons sensitive to the direction of the falling water, thus disrupting the balance between them and their antagonistic partners tuned to the opposite direction. The oppositely tuned neurons are released from inhibition and become active, which leads to the disturbing perception that the stationary rocks are moving upward! Steven Petersen, James Baker, and I showed monkeys continuously moving images, like a waterfall, and then tested the responses of their MT neurons. We found that their responses were suppressed when tested with stimuli moving in the same direction as the prior adapting movement and were enhanced for stimuli moving in the opposite direction. More recently, Roger Tootell and his colleagues have induced the waterfall illusion in humans and have found similar changes in MT monitored with functional magnetic resonance imaging (MRI).

There is additional evidence that the activity of MT neurons is directly related to the perception of motion. Kenneth Britten and his colleagues recorded from MT in monkeys that were observing ambiguous images that could be perceived as moving either in one direction or its opposite. The monkeys had been previously trained to report the direction in which they perceived motion. When the activity of the MT neuron was higher, the monkey tended to perceive the ambiguous image as moving in the preferred direction of the neuron; when the activity was lower, the monkey tended to perceive the image as moving in the antipreferred direction. Daniel Salzman and his colleagues did an analogous experiment in which they induced activity in MT neurons by stimulating them with microelectrodes. The microstimulation caused the monkey to perceive motion in the direction corresponding to the preferred direction of the neuron. Thus the activity of directionally selective MT neurons appears to cause the perception of motion in the preferred direction of the neurons.

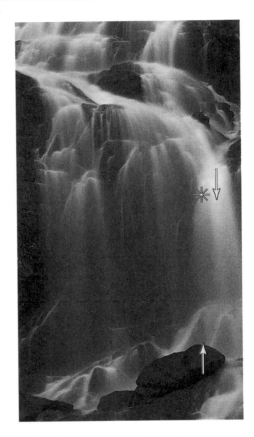

Imagine that you are gazing at a waterfall. If you were to stare at the site of the star in the midst of actually falling water for a minute and then at the rocks below, the rocks would appear to move upward in the direction opposite to that of the falling water. Be careful if you try this with a real waterfall; it can be very disorienting to see the rocks move!

## Seeing the Visual Context

The perception of qualities of objects depends heavily on the surrounding visual context. In 1982, Francis Miezin, EveLynn McGuinness, and I found that MT neurons are sensitive not only to the direction of motion of objects but also to the movement of the background. We found that when we mapped the receptive fields of MT

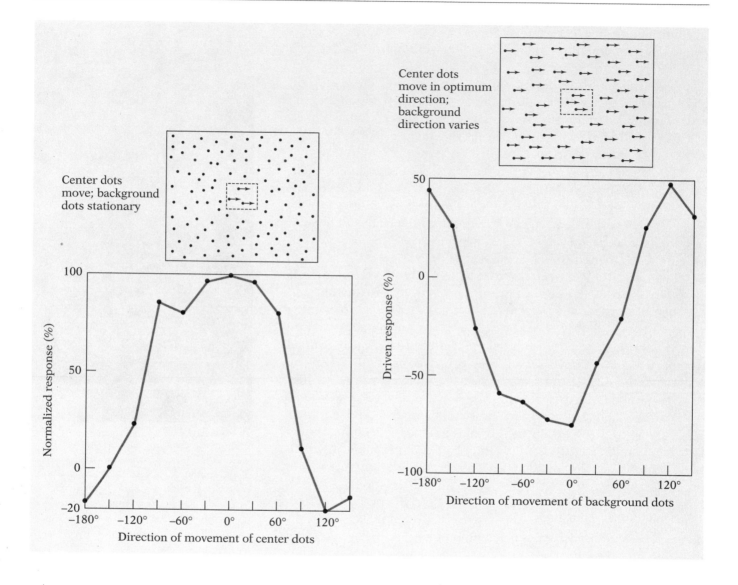

Center dots
move; background
dots stationary

Normalized response (%)

Direction of movement of center dots

Center dots
move in optimum
direction;
background
direction varies

Driven response (%)

Direction of movement of background dots

neurons on a large featureless screen, as is typically done in most vision experiments, the responses were restricted to what we called the classical receptive field. We invented this term because this field corresponds to that obtained in most visual-receptive–field mapping experiments. However, when the screen was filled with a background of coherently moving dots, we found that the direction of motion of the dots moving entirely outside the classical receptive

Opposite: The graph on the left shows how an MT neuron responded to an array of dots moving in different directions within the classical receptive field enclosed by the dotted rectangular outline. The dots in the surrounding field were stationary. The responses are plotted as percentages of the response to the best direction of movement, which was rightward (0 degrees). The graph on the right shows how the same neuron responded when the classical receptive field was stimulated with the optimum stimulus and simultaneously the direction of motion of dots in the surrounding visual field was varied. Directions of surround motion near the preferred direction suppressed the responses, and directions of motion near the antipreferred direction facilitated the responses. Thus the tuning of the nonclassical field was antagonistic to stimulation within the classical receptive field.

field had a powerful and specific effect on the responses to stimuli presented within the classical receptive field. This was very surprising, because movement of the background had no effect when there was no stimulus within the classical receptive field. Thus the response of the neuron was jointly dependent on stimuli within the classical field and outside it. Background movement in the preferred direction of movement within the classical receptive field suppressed the response, while background movement in the antipreferred direction often powerfully facilitated the response to stimuli presented in the preferred direction within the classical field.

When we mapped the sizes of the nonclassical receptive fields, we were surprised to discover that they often extended over more than half the entire visual field of the monkey. Our results indicate that the responses of cortical neurons are the product of the interaction between local cues and the global context. Analogous results have been obtained for other types of stimuli in other cortical areas. The neural tuning for object distance, described in the next section, is an example of a nonclassical effect. These effects imply that in addition to the set of dense local connections among neurons that is related to the highly ordered retinal topography of the classical receptive fields, there is a second set of connections, broader and sparser, that supports the more global responses from the nonclassical receptive field. These global effects may be responsible for many integrative aspects of visual perception such as the discrimination of figures from their background and perhaps visual memory.

## Seeing Size and Distance

Survival depends on knowing whether the furry animal in the distance is large and potentially dangerous or small and a possible meal. Determining the size and distance of objects is a fundamental feature of visual perception that probably developed early in primate evolution. More than 300 years ago, René Descartes reasoned that the perception of the size of nearby objects is related to the motor act of fixating on them, while the perception of more distant objects depends on what the viewer knows about the object and its visual context. Imagine looking at a nearby object. Your eyes converge on it. As you move the object away the angle between the lines of sight from your two eyes will decrease. More than a meter away the lines of sight will become nearly parallel and will not change very much as the object recedes farther into the distance. Thus there are large changes in the vergence angle between the eyes when fixating on objects in the near field, but small changes when they fixate on more distant objects. Similarly, the accommodative reflex causes the lens to change its optical power as a function of fixation distance, with large changes for close distances but small changes for distances of more than a meter.

In an otherwise featureless visual field that offers no clues for comparison, human subjects can discriminate the sizes of objects up to a distance of about 1 meter; at greater distances they underestimate the true sizes. This finding by Herschel Leibowitz indicates that the motor act of fixating on a object is sufficient for accurate size judgments for near objects, but that the visual context is required for judging the sizes of more distant objects. As fixation passes from the nearest possible point out to a distance of 1 meter, the vergence of the eyes and the accommodative state of the lens go through about 90 percent of their potential variation; thus beyond 1 meter there is little further variation upon which to base distance discrimination. This optical constraint means that the accurate judgment of greater distances must be based on other cues. With the full visual context available, adults can discriminate the true size of objects out to at least 30 meters, but 8-year-old children can accurately judge the sizes of objects only to a distance of 3 meters. At greater distances children underestimate the true size of objects, and the farther away the object, the greater their underestimate. Thus children seem to be unable to take full account of the visual context for distant objects because they underestimate the size of distant objects in a manner

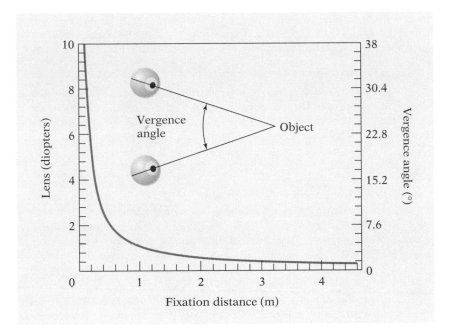

Changes in the accommodative state of the ocular lens and the vergence angle
between the eyes as a function of the fixation distance between the viewer
and the object. The curves for these two functions are identical. Note that for
distances greater than 1 meter there is little change in accommodation power
or vergence angle. This means that these cues will be of little use in determining
the distance of objects greater than 1 meter away, and that, as proposed by
Descartes, the visual system must rely on cues that are strongly dependent on
learning and experience. This is another example of how physical constraints
have influenced brain evolution.

similar to adults who cannot see the visual context. Children develop
the capacity to use the visual context to make accurate judgments by
constantly probing their spatial environment and refining their impressions. This probing proceeds in infants from the nearby space
within arm's reach and extends gradually as the child matures to
incorporate the wider world through continual feedback derived
from the experience of moving through the environment. It is easy to
forget as adults that as children we once saw the world very differ-
ently. However, this change in perception with maturation is revealed
by the common experience of returning after a long time to a place
that we occupied as children, such as a school room, and perceiving
it as adults as very much smaller than we experienced it as children.

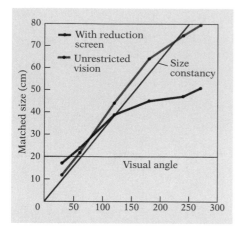

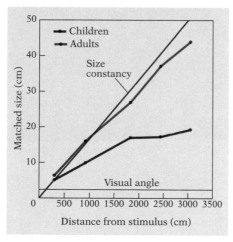

Results of experiments with the perception of object size by Herschel Leibowitz and his colleagues. Top: Adult subjects judged the sizes of objects that subtended a constant visual angle presented at different distances from the observer. With unrestricted vision the subjects were able to judge the actual sizes of the objects very accurately over the range of distances tested, and the judgments closely matched theoretical size constancy. When the visual context was eliminated by restricting the field of view with a reduction screen, the subjects underestimated the size of objects at distances greater than about 1 meter. The screen is an opaque shield with two small apertures in it. Eliminating the contextual cues to depth such as perspective and shading led to the subjects' inability to judge the distance and therefore the size of these objects. Bottom: The performance of 8-year-old children compared with that of adults. Adults slightly underestimate the sizes of objects more than 15 meters away, but children greatly underestimate the size of objects more than 3 meters distant from them. These results suggest that in judging the size of distant objects children are unable to take into account cues for distances greater than a few meters and that these cues are slowly acquired through visual experience. Leibowitz also found that as older children improve their ability to make accurate judgments of size and distance, they also begin to experience the classic distortions of size perception. Thus, paradoxically, the price of being able to see the world accurately is the susceptibility to illusions.

Accurate size judgment for near objects begins shortly after birth and depends initially on the motor acts of fixation, whereas size judgment for distant objects develops later in life and depends on experience and the ability to interpret the visual context. The contextual cues include binocular stereopsis, the cue used in stereo movies viewed through filter-glasses that allow slightly different images to be presented to each eye. Other important cues include perspective and shading, which have been used since the Renaissance by artists to represent depth in pictures. The mechanisms of vergence and accommodation probably evolved in the early primates as part of the larger set of adaptations related to high acuity vision. The slow acquisition of experience-based spatial perception is probably one of the factors responsible for the slow maturation of the brain that is characteristic of primates, about which I will have much more to say in Chapter 7.

Recently, Allan Dobbins, Richard Jeo, Josef Fiser, and I found the probable mechanism for size and distance judgments in visual cortex neurons. We tested the responses of neurons in V1 and V4 in monkeys fixating on a target on a movable monitor. We found that

This drawing of people strolling in a corridor of the Capitol shows the influence of perspective on the perception of size. The distant couple have been duplicated near the pedestal on the left. Despite the equality of the visual angles the couples subtend, the nearer couple looks very much smaller.

more than half the neurons are sensitive to differences in monitor distance and that in most of these neurons the responses progressively either decrease or increase as a function of distance between the monitor and the monkey. We called these neurons "nearness" and "farness" cells. We believe that the perception of distance results from the interaction of these two opposed populations of distance-sensitive neurons in much the same way as the direction of movement results from interaction of opposed movement detectors or, as

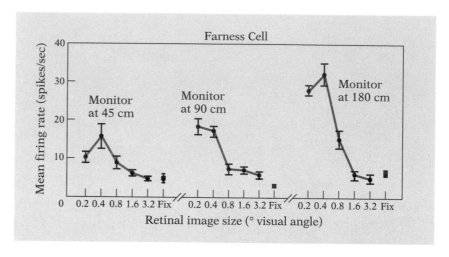

A farness cell recorded from V4. The neuron was tested with stimuli ranging in visual angle from 0.2 to 3.2 degrees with the monitor at various distances from the monkey. The value at "Fix" indicates the activity elicited by simply fixating on the screen at these distances. The recordings were made by Allan Dobbins, Richard Jeo, Joseph Fiser, and the author, from a macaque monkey that was trained to fixate on a spot on a movable computer monitor screen mounted on tracks.

I will describe shortly, the perception of color results from the interaction of opposed color mechanisms. In about half the neurons, the distance tuning was retained when we restricted the monkey's view with apertures to just the monitor screen and thus removed the surrounding visual context, the spatial layout of the room. In these neurons, the distance-related responses were probably due to vergence and accommodation. In the other half of the neurons we studied, the distance effects were probably related to cues to the visual context located in the nonclassical receptive fields of these neurons. Fixation and context combine to produce distance tuning in some cells. Visual experience is probably very important in the formation of the nonclassical receptive field properties of these neurons and the capacity to use the visual context to construct a three-dimensional world.

Our results suggest that within the two dimensional map found in each cortical visual area the third dimension is represented by the opposed populations of nearness and farness cells. Earlier theories of the visual cortex suggested that the three-dimensional world was con-

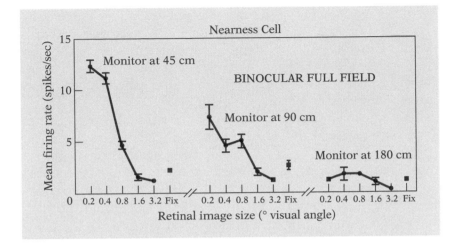

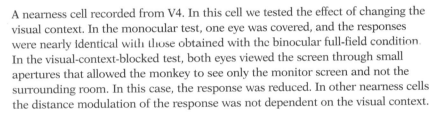

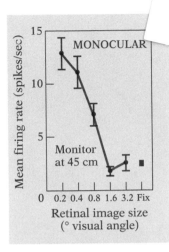

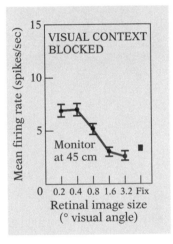

A nearness cell recorded from V4. In this cell we tested the effect of changing the visual context. In the monocular test, one eye was covered, and the responses were nearly identical with those obtained with the binocular full-field condition. In the visual-context-blocked test, both eyes viewed the screen through small apertures that allowed the monkey to see only the monitor screen and not the surrounding room. In this case, the response was reduced. In other nearness cells the distance modulation of the response was not dependent on the visual context.

structed from two-dimensional images at high levels in a hierarchy of cortical visual areas. By contrast, our results imply that distance is embedded in the maps of all cortical visual areas. However, there may be specialized areas related to three-dimensional vision within the visual cortex. One example was discovered by Russell Epstein and Nancy Kanwisher, who used functional MRI to identify an area in the medial part of the temporal lobe in humans that is selectively activated by images of the spatial layout of rooms and buildings.

## The Evolution of Color Vision

Our modern understanding of the evolution of the photoreceptor pigments comes from the beautiful experiments conducted by Jeremy Nathans in 1986, but the basic idea was a brilliant theory put forth in 1892 by Christine Ladd Franklin. She proposed that in the retinas of ancient animals there was a single photoreceptive

The theory for the evolution of color vision developed by Christine Ladd Franklin (1847–1930) was three generations ahead of its time.

pigment, and that during the course of evolution this pigment differentiated first into two and ultimately into three pigments that were sensitive to different parts of the spectrum. We now know from Nathans's work that a series of gene duplications produced these pigments; Ladd Franklin's insight anticipated the theory of evolution by gene duplication by many years.

The pigments are long chains of amino acids. Mutations in the DNA code that result in substitutions of amino acids at particular sites in the chain result in changes in the structure of the pigment molecule that alter the way the pigment responds to light from different parts of the spectrum. At about the time of the origin of vertebrates there was a gene duplication for the cone pigment that resulted eventually in the formation of two types of cone photoreceptors: one with of a pigment preferentially responsive to shorter wavelength light, toward the blue end of the spectrum, and a second with a pigment preferentially responsive to longer wavelength light, toward the red end. Animals with two cone types are called dichromats and possess a rudimentary form of color vision. Most mammals are dichromats, and the ancestor of primates was probably a dichromat. About 40 million years ago, in an ancestor of the living Old World monkeys, apes, and humans, there was a duplication of the gene for the longer wavelength cone pigment that resulted in a primates with three cone pigments. In subsequent generations in this line of primates, the spectral responses diverged in the duplicated pair of cone pigments as the result of mutations in the genes and changes in the expressed proteins. These primates are trichromats, and they perceive color in the same way normal human subjects do. The recent studies by D. Osorio, M. Vorobyev, John Mollon, and their colleagues suggest that the advance from dichromatic to trichromatic color vision specifically enhanced the capacity of primates to discriminate fruits from the background coloration of leaves.

Gerald Jacobs and his colleagues have found that in nocturnal primates, such as the galago, mutations have caused the inactivation of the short-wavelength cone pigment gene, converting these animals to monochromats. The inactivated gene, a pseudogene, is a fossil relic buried in the DNA that implies an ancestor with a greater capacity to discriminate color. It is an intriguing possibility that the DNA may contain many such fossil genes that may tell us much about the structure and behavior of ancestral forms. The so-called junk DNA, which does not encode proteins, may be a rich source of such fossils waiting to be uncovered.

The remaining cone type in nocturnal primates may serve to regulate daily activity rhythms. For example, Robert Martin has observed that galagos become active at sunset at just the time when it is difficult to discriminate color because of decreased light. This observation suggests that when cone activity ceases in the retina, a signal is relayed to the brain's clock in the hypothalamus that in turn causes the galago to become active. I will discuss this clock further in Chapter 7.

Stephen Polyak proposed that the capacity in animals to perceive color co-evolved with the capacity of plants to produce brightly colored flowers and fruits. The flowering plants, the angiosperms, first appeared in the Cretaceous period about 120 million years ago, and their great success parallels the rise of mammals. The plants evolved to produce fruit and flowers that appealed to our ancestors, who served as agents for the plants' reproduction. Some primates are nectar eaters, and thus they could pollinate flowers in much the same way as do bees. Nearly all primates eat fruit to some extent, and thus they could disseminate those seeds that passed intact through their digestive tracts. They even provided a lump of organic fertilizer to nourish the growth of the seed on the forest floor. Thus primates have served as agents of selection in the evolution of tropical rain forest plants, and this is perhaps why the appearance, odor, and taste of fruit and flowers are so attractive to us.

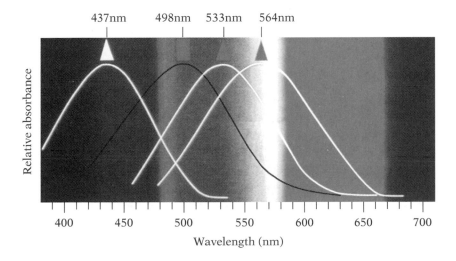

The spectral tuning of the three retinal cone pigments is shown by the white curves, that of the rod pigment by the black curve. The peak sensitivity is shown in wavelengths. The 564-nm pigment and the 533-nm pigment resulted from a gene duplication that occurred about 40 million years ago in the ancestor of Old World monkeys, apes, and humans. The amino-acid sequences for these two cone pigments are 96 percent the same, reflecting the relatively recent divergence of their genes. The cone pigments are particularly striking examples of the divergence of function following gene duplication.

The usefulness of color vision for detecting ripe fruit. Someone with a lesion in cortical area V8 in the right hemisphere, gazing at the place indicated by the star in this array of ripe peaches, would see the right half of the scene in full color and the left half in shades of gray, as shown here.

Primates have evolved neural circuits, specializations of the neurons in the parvocellular pathway, for the analysis of the input from the different types of cones. The hallmark of color circuits are the spectrally opponent neurons. For example, some are excited by green light and inhibited by red light, others are excited by blue light and inhibited by yellow. Studies done by Russell DeValois, Torsten Wiesel, David Hubel, and Margaret Livingstone found that spectral opponency is a characteristic feature of neurons in the parvocellular layers of the lateral geniculate nucleus, and in parts of V1, the primary visual area.

One of the color-sensitive structures in V1 is a set of regularly spaced spots, prosaically named "blobs," located in the upper cortical layers. However, the parvocellular layers of the lateral geniculate nucleus and the blobs are present and well developed in nocturnal species, suggesting that color processing is not the only thing these structures do. The blobs project to a series of stripes in V2, which in turn project to two higher visual areas, V4 and VP (the ventral posterior area). V4 was once considered to be the "color area" and was thought to be crucial for the perception of color, but careful studies by Peter Schiller and by Alan Cowey and his colleagues have shown that damage to V4 produces only mild deficits in the perception of color. By contrast, Schiller found that the same V4 lesions produced severe deficits in the ability to judge the size of objects. In both humans and monkeys, there exists an area beyond V4 and VP, located on the ventromedial side of the occipital lobe, that when damaged by a stroke or other injury results in a dramatic loss of color vision in the affected part of the visual field. This area, called V8, has recently been mapped in humans with functional MRI.

## Making and Seeing Faces

Primates rely on facial expression to communicate their emotions. In the evolution of primates, the importance of facial expression expanded while the olfactory system regressed. Specialized scent glands for the communication of social signals and the olfactory system for the reception of these chemical signals are well developed in prosimian primates. In monkeys, apes, and humans, however, the olfactory system is much reduced, and some of its functions have been taken over by the visual perception of facial expression. EveLynn McGuinness, David Sivertsen, and I found that the muscles

## Color in V8

In 1888, the French neurologist D. Verrey described a patient who had lost the capacity to perceive color throughout half the visual field. Subsequently the patient died, and the autopsy revealed a well-defined lesion in the ventral occipital lobe on the side of the brain opposite to the color-blind half-field. Similar cases of acquired color blindness, or achromatopsia, associated with lesions in the same locality have been reported a number of times. One of the clearest clinical accounts was provided in 1980 by Antonio Damasio and his colleagues, who found that their patient "was unable to recognize or name any color in any portion of the left field of either eye, including bright reds, blues, greens and yellows. As soon as any portion of the colored object such as a large red flashlight was held so that it was bisected by the vertical meridian [midline], he reported that the hue of the right half appeared normal while the left half was gray. Except for the achromatopsia, he noted no other disturbances in the appearance of objects [i.e., objects did not change in size or shape, did not move abnormally, and appeared in correct perspective]. Depth perception in the colorless field was normal. The patient had 20/20 acuity in each eye." Brain-imaging revealed that the patient had a lesion in the right occipital lobe in the same position as in Verrey's patient. Damasio noted that the loss of color perception throughout the entire hemifield implied that the entire opposite half of the visual field was mapped in a color area in the ventral part of the occipital-temporal lobe. This ran contrary to the conventional wisdom of the time, which held that only the upper visual field was represented in the ventral part of the brain. However, this is precisely what Roger Tootell and his colleagues found when they mapped the area by presenting colored stimuli and monitoring the responses with functional MRI.

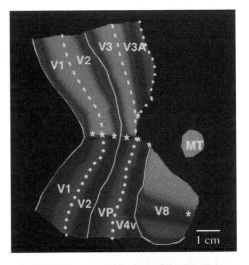

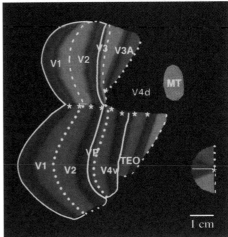

Top: An unfolded map of the human visual cortex based on the functional MRI studies of Roger Tootell and his colleagues, showing visual area V8 relative to other cortical visual areas. The map is not complete. Additional areas link MT to the mapped regions. Bottom: A comparable map for the visual areas in the macaque monkey based on neurophysiological recording. Note that the area corresponding to the human V8 is TEO, the tempero-occipital area. The color coding for polar angle is shown in the inset visual field map in the lower right corner.

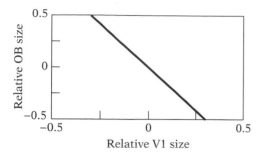

As visual structures expanded during the course of evolution, olfactory structures diminished in size. This tradeoff between olfaction and vision reflects an increased reliance on visual social communication via facial expressions and gestures and a decreased use of olfactory signals. An example of this tradeoff is illustrated in this graph, which shows that the olfactory bulb (OB) declines in size as the primary visual cortex (V1) expands when different primate species are compared. To construct this graph, Andrea Hasenstaub and I used the brain volume and body weight data of Heinz Stephan and his colleagues and the residual method for calculating relative size, which is described on page 165.

that produce facial expressions are very well represented in the motor cortex of monkeys. We found a large amount of cortex devoted to the muscles that retract the corners of the mouth and smooth or wrinkle the skin around the eyes, which are important in producing the expressions of fear, threat, and play. The large cortical representation of these muscles suggest that facial expressions are not mere automatic responses to behavioral states but are under some degree of conscious control. Thus higher primates are all to some extent actors manipulating their expressions of emotion within a social context. The much greater range of emotional expression and subtlety of control afforded by the facial muscles is perhaps why facial expressions have tended to supplant involuntary olfactory social cues in primate evolution.

In Chapter 2, I described the amygdala as a part of the forebrain that is involved in the control of social behavior. The amygdala receives inputs from both the main olfactory and vomeronasal systems and in turn influences the hormonal systems through the hypothalamus. It has also been implicated as a key part of the mechanisms for learning fearful responses. The amygdala also receives input from the inferotemporal visual cortex, which is greatly expanded in higher primates. Charles Gross, Robert Desimone, Edmund Rolls, and David Perrett and their colleagues have shown that neurons in parts of the inferotemporal cortex are especially sensitive to the images of faces.

The role of the amygdala in the perception of facial expressions was beautifully shown by Ralph Adolphs and his colleagues, who studied a remarkable patient who had suffered bilateral amygdalar damage without significant injury to other parts of the brain. Although this patient had normal vision and could perceive faces, she was unable to discriminate the emotional content in the negative facial expressions of fear and anger. Thus all faces appeared to be smiling or neutral to her, even those which were actually frightened or angry. In functional brain-imaging studies, J. S. Morris, Paul Whalen, and their colleagues found that normal subjects who saw fearful faces had increased activity in the amygdala but decreased activity when they saw happy faces. However, the amygdala's role is not limited to visually communicated emotional states. Sophie Scott and her colleagues found that amygdalar lesions also disrupted the ability to perceive the emotional content of speech intonation even though their patient had normal hearing. As with facial expressions in Adolphs's patient, the auditory expressions of fear and anger were the most impaired in this patient.

Facial expressions in macaque monkeys, drawn from life by Leslie Wolcott. Clockwise from top left: A relaxed expression while grooming; a play face; an expression of aggression; intense fear; mild apprehension. The macaque at the center is stuffing a banana into her cheek pouches.

## Establishing Priorities

One of the fundamental problems faced by all nervous systems is how to sort from the immense flood of incoming information which bits are important and which can be safely ignored. Compared with the vast and sometimes conflicting sensory input, the set of behavioral choices available to the organism at any one moment are very much smaller. Thus the brain must establish priorities regarding

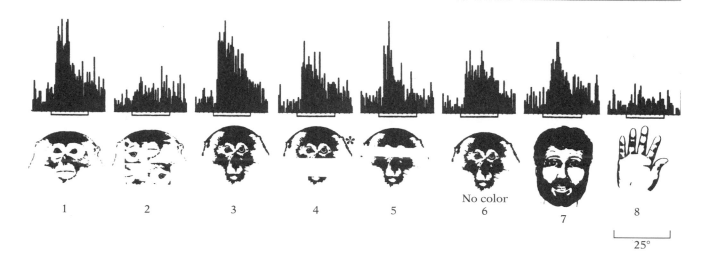

A face cell recorded from the inferotemporal visual cortex in a macaque monkey by Robert Desimone and his colleagues. Each histogram illustrates the cell's response; the stimulus was on during the underscored period. 1, strong response to the image of a monkey face; 2, little or no response to a scrambled image of the same monkey; 3, strong response to another monkey face; 4, less response when the mouth was covered; 5, less response when the eyes were covered; 6, slightly less response to the face without color; 7, response to the face of one of the investigators; 8, lack of response to the image of a hand.

perceptions and thoughts since they are not all equally significant, nor can they all be acted upon at once. Similarly, the brain must establish priorities in the sequential timing of behavioral responses, because some acts require the successful completion of prior actions. In primates, the frontal lobe has an important role in establishing priorities and planning. In particular, the lower surface of the frontal lobe, termed the orbital-frontal cortex, is especially important for these functions, as has been shown by an extraordinary series of clinical observations of brain-damaged patients by Antonio Damasio and his team in the Department of Neurology at the University of Iowa College of Medicine. One of these patients presents a particularly poignant example.

Patient E. was a corporation executive with a high income, a high IQ, and a good marriage, a man highly respected by his family and the very model of a successful person. A tumor, formed in the membranous lining of his brain, was surgically removed together with the adjacent orbital-frontal cortex. Following the surgery he quickly regained his health and there was no reduction in his IQ score, yet his life utterly fell apart. He lost his executive position; his marriage collapsed; he made foolish financial decisions and soon went bankrupt. He realized that something horrible had happened as the result of his tumor and the surgery. He sought public assistance, but no one would believe that someone who had been so successful, who was obviously highly intelligent, and who appeared perfectly normal in other respects, needed aid from the state.

E.'s problem was that he could not establish priorities among different options. He was paralyzed in his capacity to make the simple choices that fill everyday lives. What to do next? Which articles of clothing to wear? What to eat? In each instance his mind was filled with alternative courses of action that he could neither accept nor reject. In attempting to choose a restaurant his mind was flooded with data about the multitude of advantages and disadvantages of many different places to dine. The food was good at one, but the service was slow, or the service fine, but the noise level unpleasant. In the end he simply could not make up his mind. His thinking was logical but unchecked by any realization that some factors are more important than others. His excellent memory contributed to his impairment because it provided so much material upon which to base his indecision. His deficit made it impossible for him to hold on to any sort of stable employment or relationship.

E. and other patients with orbital-frontal cortex damage are deficient in their physiological response to emotion-laden stimuli. Damasio and his colleagues tested this by measuring changes in the electrical conductivity of the skin resulting from small increases in perspiration associated with arousal when subjects were shown images depicting violence or sexual situations. Normal subjects are strongly aroused by these stimuli, but the orbital-frontal patients showed no reaction, although they did respond to sudden loud noises, indicating that some of the circuitry responsible for the skin response was intact. During the testing, E. commented on his detachment from the emotional content of the images, saying that before his surgery the images would have been exciting to him but were no longer so. Damasio and his team also tested E. and the other orbital-frontal patients on a gambling task and found that here too they differed markedly from normal subjects. In this game, the subjects were presented with different stacks of cards. Each card carried a monetary reward or penalty. Some stacks had cards with small rewards and small penalties, but the aggregate was positive. Other stacks had cards with large rewards and even larger penalties, and the aggregate was very negative. Normal subjects quickly learned to choose cards from the stacks that were generally positive, but the orbital-frontal patients fixed on the high reward stacks even after receiving many costly penalties that deterred normal subjects from choosing them. Damasio theorizes that the decision-making processes depend on signals arising in the orbital-frontal cortex that can either generate gut feelings overtly or, instead, influence decision-making covertly

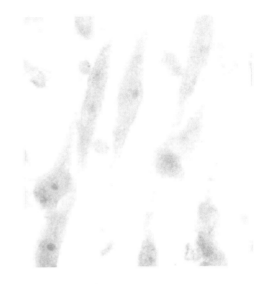

The anterior cingulate cortex lies just medial to the orbital frontal cortex and is closely connected with it. Recently, Patrick Hof and I and our colleagues found a unique population of neurons in layer 5 of the anterior cingulate cortex of humans and their closest relatives, the great apes (bonobos, chimpanzees, gorillas, and orangutans). These large spindle-shaped cells, shown above, possess both an apical dendrite extending toward the cortical surface and a single basal dendrite extending toward the white matter. Spindle cells have been found only in humans and the great apes, but not in other mammals after an exhaustive search. While not quite a "hippopotamus major," this feature does constitute a clear example of an anatomical specialization in the brains of humans and their close relatives that probably reflects a difference in cortical circuitry from other mammals. Brain-imaging and clinical studies in humans suggest that the functions of the anterior cingulate include social awareness, self-control, and the capacity to perform tasks requiring intense cognitive effort. The spindle cells are especially vulnerable to degeneration in Alzheimer's disease.

by introducing biases in the reasoning process. When normal subjects experience situations that they associate with the possibility of negative consequences, they begin to have a more rapid heart rate and to have changes in the peristaltic action of the gut. They sense their gut feelings and make decisions accordingly. Damage to the orbital-frontal cortex disrupts the capacity to make these physiological responses to situations that signal an unfavorable outcome for the individual, and thus the ability to make crucial survival decisions is severely compromised.

## Unique to Primates:
## A Center for Visuo-Motor Coordination

The visual guidance of body movements is particularly important in primates. Both theories for the origin of primates, the visual-predation and the fine-branch-niche hypotheses, stress the importance of visuo-motor coordination. In an extensive study of the cortical sites that connect to the spinal cord in 22 species of mammals, Randolph Nudo and Bruce Masterton found a cortical site projecting to the spinal cord that is unique to primates, the ventral premotor area. Thus in primates an additional cortical area has evolved for the control of the muscles of the body. Giacomo Rizzolatti and his colleagues found that neurons in the ventral premotor area are activated when the subject performs visually guided reaching and grasping movements such as when the monkey manipulates objects. Neurons in this area also respond when the subject observes the experimenter performing the same task. Because of this property, these neurons have been called "mirror" cells.

The mirror cells suggest that this unique primate motor area may also be involved in observational learning of visually guided tasks. Broca's area, the region of the cortex involved in the production of speech sounds, is located in approximately the same position in the human brain as the ventral premotor area in other primates. Broca's area may be a specialization associated the ventral premotor area, a possibility I will discuss in Chapter 7.

When Nudo and Masterton compared the amount of cortex that projects to the spinal cord relative to the total amount of cortex in different primate species, they made the interesting discovery that the size of the projecting cortical areas always constituted a con-

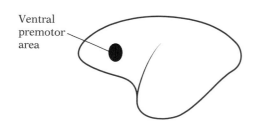

Ventral premotor area

The position of the ventral premotor area, a cortical structure unique to primates, from the work of Randolph Nudo and Bruce Masterton.

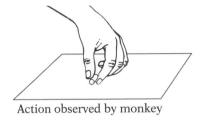

Action observed by monkey

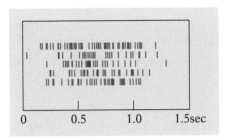

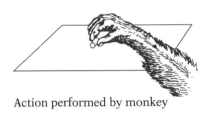

Action performed by monkey

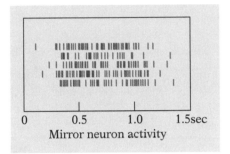

Mirror neuron activity

A mirror cell recorded from the ventral premotor area in a macaque monkey by Giacomo Rizzolatti and his colleagues. The spike trains at the right show the activity of the neuron during a series of repeated actions, by the monkey itself (bottom) and by the human investigator (top). There are similar mirror neurons for facial movements.

stant fraction of the total cortex. Thus the total amount of cortex devoted to motor output expands in proportion to the size of the entire cortex. This finding suggests that the refinement of motor control is proportional to the processing power of the cortex as a whole.

## Saving Wire: The Formation of Cortical Maps and Fissures

The visual field is represented in two different kinds of maps in the visual cortex in primates. As in the optic tectum in primates, all the maps are of the opposite half of the visual field. First-order transformations of the hemifield are topologically like the tectal map. In this type, the representation of the central retina stretches the map but there are no cuts in the map of the hemifield. The maps in the primary visual cortex and area MT are first-order transformations. Second-order transformations contain maps in which the representation of the hemifield is split along the representation of the

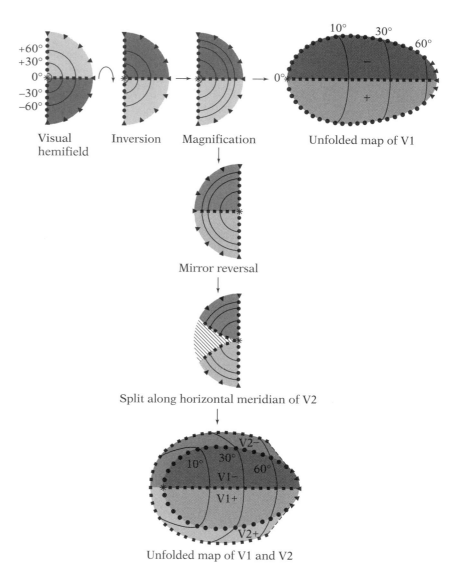

The transformation of the visual field map in the primate visual system
in different stages. The visual hemifield is inverted by the lens and the
representation of the central visual field is magnified in stages between the
retina and primary visual cortex (V1). In V2, the map is the mirror image
of V1; it is split along the representation of the horizontal meridian so that
it can wrap around the vertical meridian representation of V1. The maps of
the visual hemifield in V1 and V2 are not topologically equivalent because
of the split horizontal meridian in V2.

horizontal meridian. These are nontopological maps because adjacent points in the hemifield do not necessarily map onto adjacent points in the cortex. The representations in V2 and many other visual areas fall into this category. The second-order transformations allow the different maps to fit together in the cortical sheet in such a way that there are no major discontinuities in the visual field representation at the junctures between areas. This form of mapping reduces the length of fiber connections. The densest connections are among representations of adjacent parts of the visual field, and this continuous mapping serves to minimize the distance traversed by fibers linking these adjacent parts, but at the expense of a few much longer connections near the split parts of the maps. Maps also have reciprocal connections between the same parts of the visual field. For example, V1 and V2 are reciprocally connected, and in this case the split representation of the horizontal meridian in V2 appears to shorten the path length of these reciprocal connections.

Shortly after Jon Kaas and I proposed this theory, Bruce Dow suggested to me that the same idea might explain why in monkeys with large brains portions of V2 fold back upon V1. This folding allows the representations of the same parts of the visual field in V1 and V2 to lie adjacent to each other and minimizes the distance traversed by reciprocal connections between V1 and V2. More recently, David Van Essen has proposed that the mechanical tension produced by short-wire connections between areas actually pulls the two areas together as they grow in the developing brain, thus causing folds to form in the cortical sheet. Van Essen's "pulling strings" theory for how cortical folds develop has not yet been tested, but is an interesting possibility. The theory that wire length is minimized is supported by a quantitative analysis of the connections of the visual cortex in macaque monkeys. The principle of minimizing wire length appears to be a general factor governing the connections of nervous systems. Christopher Cherniak showed in the well-mapped nervous system of nematode worms that the shortest possible pathlengths are used to connect the neurons. Cherniak calculated the total length of millions of possible wiring patterns that might be used to connect the components of the nematode nervous system and found that the one actually used is the most economical in terms of the length of connections. Thus natural selection strongly favors wiring economy. The same design principle applies to brains made of silicon. Carver Mead has emphasized that one of the main

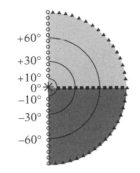

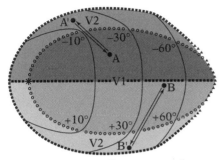

Actual Topologically Nonequivalent V2

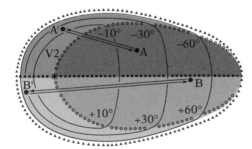

Hypothetical Topologically Equivalent V2

The actual topologically nonequivalent V2 compared with a hypothetical topologically equivalent V2. Sites in V1 and V2 that represent the same location in the visual field are reciprocally connected. The distance traversed by these reciprocal connections would be much greater in the topologically equivalent V2.

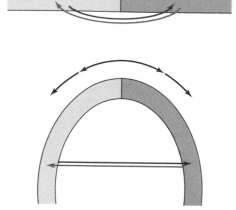

David Van Essen's string theory of cortical fold formation. The blue and the tan areas are reciprocally connected. As the cortical areas expand (as indicated by the arrows), the connections force the cortex to fold. Note, however, that the amount of space required for neocortical connections in the white matter increases disproportionally with respect to the expanding neocortex, as discussed on page 116.

constraints in the design of very large integrated circuits on silicon chips is the need to minimize the length of wire required to link functional components.

## The Evolution of Multiple Cortical Areas

The neocortex in primates comprises 50 to 100 distinct areas as defined by functional and anatomical criteria. This number will certainly grow as our knowledge of the functional anatomy of the neocortex expands. The idea that the neocortex is made up of areas that perform unique functions is centuries old and has had an enormous appeal to scientists and nonscientists alike. Many areas appear to share common organizational features that suggest that they arose through duplications of pre-existing areas: prime examples are the first and second visual areas, V1 and V2. They could have come about as the result of mutations that caused whole areas to be replicated with the divergence in function coming in subsequent generations as Jon Kaas and I proposed many years ago. Unfortunately, there is still no good test of this theory, although, as discussed in Chapter 5, the discovery by Cohen-Tannoudji and his colleagues of a gene linked closely to the development of the primary somatosensory cortex suggests that this theory is still a possibility.

Leah Krubitzer has proposed that modular substructures within areas, such as the cytochrome oxidase blobs within V1, represent a stage in the formation of new cortical areas. However, the alternative possibility is that the modules may simply represent efficient ways to embed multiple subsystems within a single cortical map. This alternative is supported in the case of the blobs by evidence that they have been remarkably stable structures in evolution. The blob–interblob architecture is present in V1 without exception in New World monkeys, Old World monkeys, apes, and humans, implying that it was present in the common ancestor of these primates which lived more than 45 million years ago. The long-term stability of the blob system in V1 contrasts with the apparent timing of the origin of the visual areas beyond V2 in primates. We do not know long it took for the areas beyond V2 to emerge, but a reasonable inference from the available data is that they took shape between 55 million and 50 million years ago. There is no fossil evidence for large, frontally directed eyes and an expanded visual cortex before the Eocene period, which began 55 million years ago. The origin of these areas probably occurred during the expansion of occipital and temporal cortex that can be seen in the endocasts of the primates living in the early Eocene. Further evidence that these areas were present in the early primates is the existence of many of them, such as MT, V4, and the inferotemporal cortex, in both prosimian and simian species, indicating that they were present in their common ancestor, which would have lived at about this time. Taken together, these lines of evidence suggest that the origin of many of the visual areas beyond V2 may have occurred over the span of only a few million years. While I am skeptical that the blob system in V1 represents a stage in the process of the fission of cortical areas, nevertheless Krubitzer's hypothesis for the formation of new cortical areas is an interesting possibility that could be examined through experimental manipulations of cortical inputs and computer simulations of cortical architecture.

The idea that each cortical area has a distinctive function is an attractive notion, which is well supported by the specialization for motion perception in MT and color perception in V8. However, evolution through gene duplication suggests alternative ways of thinking about the functional roles of the different areas. The duplicated genes for the cone pigments encode photoreceptor proteins that have diverged in their responsiveness to different parts of the spectrum. Each protein continues to be responsive to all parts of the visible spectrum; the differences enable the animal to perceive color, but each by itself is insufficient to sustain color vision. Another example is the inner and outer hairs cells, which cooperate to achieve improved hearing in mammals. Improved perceptual functions emerge from the combination of inputs from cone pigments and from the cooperative interactions of inner and outer hair cells, and so it may for cooperative interactions among cortical areas as well. Still another example is the hemoglobin molecule that is made up of four protein chains that are the result of a fourfold replication of the gene for primordial chain early in vertebrate evolution. Collectively the four chains bind and release oxygen more efficiently than does the single chain variant. These examples indicate that the products of replication events need not be functionally independent but rather achieve their evolutionary utility through their cooperative interactions. The same is likely to be true of cortical areas. With this idea in mind, I will turn in the next chapter to the evolution of the brain as a whole.

Cast adrift in the Tiber by a wicked uncle, the royal twins Romulus and Remus were rescued, suckled, and nurtured by a she-wolf. They survived, in later years to reclaim their heritage and build a new city, Rome, near the place they had washed ashore. This sculpture of the wolf, in the Capitoline Museum of Rome, is Etruscan, c. 500 B.C.; the twins are a Renaissance addition. Turkish, Persian, Teutonic, Irish, Aztec, and Navaho legends also have accounts of wolves that nurtured humans, all possibly deriving from early observations of the close family ties within wolf societies.

# The Evolution of Big Brains

Strong social bonds, high levels of intelligence, intense parenting, and long periods of learning are among factors used by higher primates to depress environmentally induced mortality. It is of some interest that such factors also require greater longevity (for brain development, learning, acquisition of social and parenting skills) and that they constitute reciprocal links leading to greater longevity.

Owen Lovejoy,
*The Origin of Man*, 1981

Animals with big brains are rare. If brains enable animals to adapt to changing environments, why is it that so few animals have large brains? The reason is that big brains are very expensive, costly in terms of time, energy, and anatomical complexity. Large brains take a long time to mature, and consequently large-brained animals are dependent on their parents for a long time. The slow development of large-brained offspring and the extra energy required to support them reduce the reproductive potential of the parents. Thus extra-special care must be provided to insure that the reduced number of offspring survive to reproductive age. Brains must also compete with other organs for energy, which further constrains the evolution of large brains. Finally, as discussed in Chapter 5, the amount of brain devoted to wiring connections tends to increase disproportionally with brain size, imposing an additional barrier to the evolution of large brains. The basic question is, how do those few animals with large brains bear these extra costs?

## Bodies, Brains, and Energy

Primates tend to have larger brains than other mammals, but even among primates there is large variation in brain size in the different species. Obviously, part of this variation is related to differences in body weight because brain weight scales with body weight. Many other things scale with body weight, such as the power consumption of the body, metabolic rates, the time required to reach a particular developmental stage, such as sexual maturity, and even life span. A simple equation expresses this general relationship:

$$Y = kX^a$$

where $Y$ is brain weight, the constant k is the scaling factor, $X$ is body weight, and a is an exponent.

Because the values of $X$ and $Y$ vary over several orders of magnitude, it is convenient to transform them into logarithmic scales, so that the equation becomes

$$\log Y = a \log X + \log k$$

For primates, the equation becomes

$$\log \text{brain weight} = 0.75 \log \text{body weight} - 0.94$$

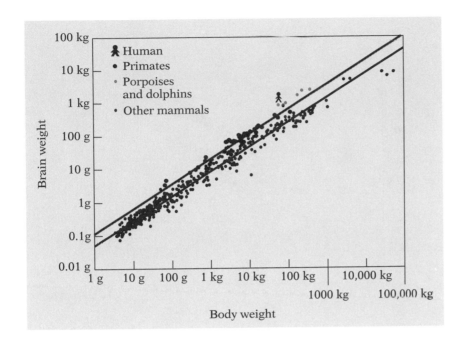

Brain weight

Body weight

- ♣ Human
- • Primates
- • Porpoises and dolphins
- • Other mammals

Primates have larger brains for their body weight than do most other mammals, a relation expressed by the regression line for primates, which is nearly parallel to the lower regression line for nonprimates. On a log–log scale, power-law relationships such as the allometric equation relating brain and body weight plot as straight lines. Note, however, that the mammals nearest to humans in terms of the brain–body relationship are porpoises and dolphins, which also have large brains for their bodies. The data used to construct this graph were generously provided by Robert Martin.

For nonprimates, the equation becomes

$$\text{log brain weight} = 0.74\ \text{log body weight} - 1.3$$

Thus primate brains scale with body weight with almost exactly the same exponent value (0.75) as for the nonprimates (0.74), which is also expressed by the nearly parallel regression lines in the log brain–log body plot. Thus primate brains tend to increase at nearly the *same rate* as a function of body weight as do those of nonprimates. However, primate brains tend to be about 2.3 times larger than the brains of nonprimates of the same body weight, and this is expressed by the difference in the scaling factor.

The same is true of fetal brains. For any size fetus, the brain tends to be about twice as large in a primate as in a nonprimate. The only exceptions are the toothed whales, which are intermediate between primates and nonprimates. Thus, Robert Martin concluded that "at any stage of fetal development, primates devote a greater proportion of available resources to brain tissue than do any other mammals."

A logarithmic plot of fetal brain weight against body weight for a sample of mammalian species, from the work of Robert Martin. Note that for every fetal size primate brains are larger than those of nonprimates.

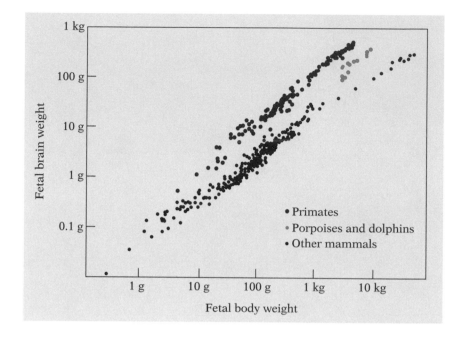

It has often been suggested that brains scale with the surface area of the body because it is through its body surface that an animal interacts with its environment. This theory predicts that the brain should scale to the 2/3 power of body weight since body surface area scales with an exponent of 2 (square units of area) and body weight with an exponent of 3 (cubic units of volume). As Martin has pointed out, this theory is not supported by data from primates or for the whole group of mammals, where brain weight scales very closely to the 3/4 power of body mass. Furthermore, only a small fraction of the brain is involved in the processing of the sensory input from the body surface, and within these somatosensory structures, the representations of very small parts of the body surface, the hands and tongue, for example, predominate. In primates, a large fraction of the brain is devoted to processing the input from the central retina, which amounts to less than 1 percent of the retinal surface.

The rate of energy use by animals at rest, measured in watts, also scales at the 3/4 power of body weight. Thus brain weight increases as a function of body weight just as the energy requirement does. Geoffrey West and his colleagues have shown that the 3/4 power

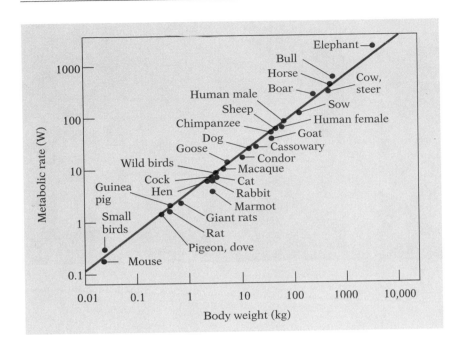

The relationship between resting metabolic rate measured in watts and body weight for mammals and birds. Energy consumption increases at the 0.75 power of body weight.

relationship between body weight and metabolism is the result of the branching geometry of the blood vessels that nourish the body, and it is likely that related geometrical constraints will ultimately prove to be responsible for brain scaling as well.

In primates with the same body weight, there is considerable variation in brain weight. For example, humans and chimpanzees are about the same body weight, but the human brain is 3 times larger than the ape's. It is useful to compare the relative sizes of the brains of animals with different body sizes. However, it is misleading to make this comparison by simple ratios because of the power-law relationship between brain and body weight. To measure differences in relative brain size for animals with different body weights, it is necessary first to calculate a linear regression to determine how brain weight varies as a function of body weight for the whole sample of primate species. The regression line provides an estimate of how large the brain would be in a typical primate of a given body mass. The concept of the regression line for illustrating the average tendency within a population described by two variables was invented by Darwin's cousin, Francis Galton, in 1885. The distance

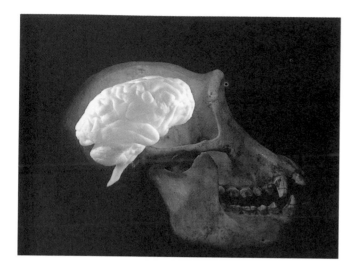

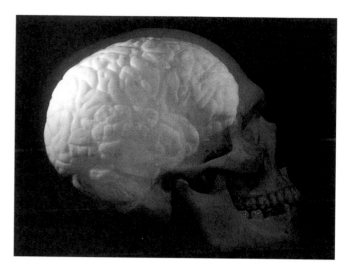

The relative sizes of chimpanzee and human brains.

between the regression line and the data point for each species is a measure of how much variation remains for that species after the effect of body weight has been removed. This is the residual variation for each species, and it can be either positive or negative with respect to the regression line. The residual variation is thus the relative brain size for that species after the effect of body mass has been removed. This residual variation is a measure of differences in brain evolution within a closely related group of animals. The fundamental question is what factors are related to this residual variation in brain size.

A quick inspection of the brain–body weight graph reveals that primate species that eat mainly fruit tend to have larger brains than do primate species that eat mainly leaves. This impression is confirmed by measuring the residual variation for fruit eaters versus leaf-eaters. Fruit-eating primates do have significantly larger brains than leaf-eaters. Frugivory is also linked to relatively larger brain size in other groups of animals. For example, fruit-eating bats have larger brains than do other bats, and parrots, which eat predominately fruit and nuts, have larger brains than do other birds. The association between frugivory and large brain size is probably linked to the special problems of harvesting fruit in the complex mixture of different types of trees in the tropical rain forest. Fruit trees are widely dispersed in the rain forest and bear fruit at different times. The spatial and temporal dispersion of fruit resources presents the

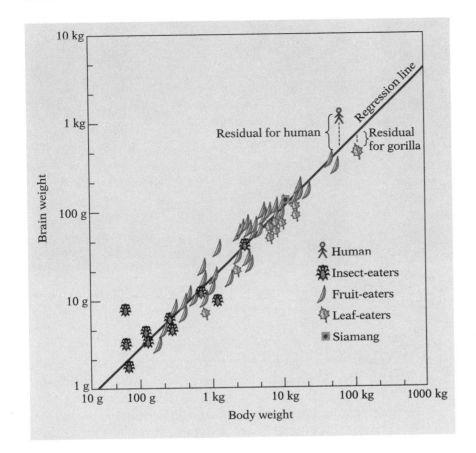

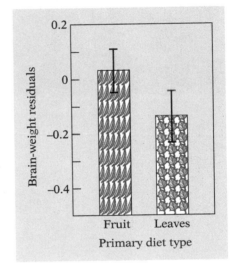

forager with complex problems. The successful harvesting of ripe, digestible fruit requires that the animal remember the location of potentially fruit-bearing trees and anticipate when they will be in season. In addition, because of the high nutritional quality of ripe fruit and its easy digestibility, there is often intense competition from other animals for these scarce resources. Because of these factors fruit-eaters must *plan* their foraging expeditions carefully if they are to survive. By contrast, leaves are ubiquitous and can be easily harvested with little competition for these abundant resources. Ripe fruits are a much rarer and more variable resource in space and time than are leaves. The existence of larger brains in fruit-eaters than in leaf-eaters supports the hypothesis that the brain helps the animal to cope with environmental variation. It is also consistent with evidence discussed in Chapter 1 and Chapter 3 that the most basic function of

The relation between brain and body weight for primate species. The residual variance in brain weight after removing the effect of body weight is calculated by measuring the vertical distance between the data point for each species and the regression line. The graph on the right, based on the brain residuals, shows that primates that eat mainly fruit have significantly heavier brains than do those eating mainly leaves.

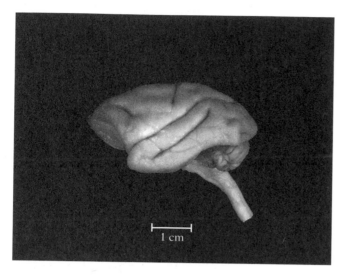

Howler monkey

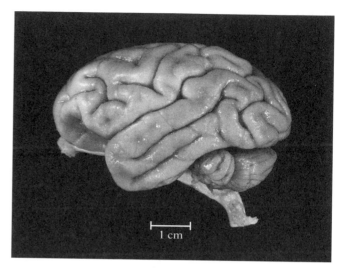

Spider monkey

A comparison of the brains of two closely related New World monkeys with about the same body weight. The smaller-brained howler monkey eats predominately leaves; the larger-brained spider monkey is a fruit-eater. At 54 grams the howler monkey brain is only one half the weight of the spider monkey brain (108 grams). Note also that the spider monkey cortex has many more fissures than the howler monkey cortex. The brains are from Wally Welker's collection, which can be accessed at http://www.neurophys.wisc.edu/brain.

the brain is to control what is taken into the gut. Indeed the gene *BF-1*, which regulates forebrain size, is closely related to genes controlling the formation of the gut.

The difference in brain size between fruit-eaters and leaf-eaters is also linked to differences in energy budgets. Leslie Aiello and Peter Wheeler have pointed out that the energy costs of digesting leaves are much greater than for digesting ripe fruit. These costs are incurred because the complex carbohydrates in leaves must be fermented and broken down into usable simple sugars, a process requiring a large specialized gut (stomach and intestines) and a considerable amount of energy. Aiello and Wheller found that the size of the digestive organs is negatively correlated with brain size. The brain tends to expand at the expense of the digestive organs, and vice versa. This trade-off exists because the overall energy use by an animal is a function of its body weight. With body weight determining the sum total of watts available to an animal for all its bodily functions, it follows that if the energy devoted to one organ is greatly increased, then there must be a commensurate decrease in the energy used by the other organs. The main energy-using organs are the heart, liver, kidney, stomach, intestines, and brain. The sizes of the heart, liver, and kidney are very tightly linked to body mass, forcing a trade-off between brain and digestive organs. These differences in gut and brain size result from a trade-off in the

animal's energy budget between watts devoted to digestion and watts devoted to brain metabolism. On the one hand, the animal may use a larger portion of its quota of watts to digest abundant, easily harvested foods (leaves), but these foods contain nutrients that are relatively inaccessible biochemically and thus costly to digest. On the other hand, the animal may use a larger portion of the available watts to support an enlarged brain with the capacity to store information about the location and cognitive strategies necessary to harvest food that is scarce and hard to find (ripe fruit), but these foods are less costly to digest. Leaf-eaters have an additional burden because leaves very often contain toxins which require further energy expenditures to detoxify.

A howler monkey chewing leaves; a spider monkey eating fruit.

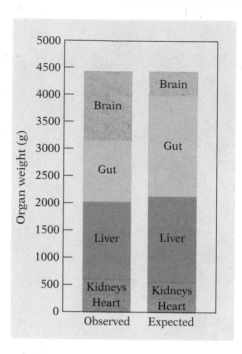

The observed and expected organ weights for a standard 65-kilogram human, from the work of Leslie Aiello and Peter Wheeler. The expected values for the human were calculated from regressions based on the data for these organs relative to body weight for primate species. Note that the human brain is 800 grams larger than would be expected for a 65-kilogram primate and the gut is about 800 grams smaller.

The evolution of improved color vision may have facilitated the evolution of larger brain size in primates. Fruit color indicates the degree of ripeness and therefore the nutritional content and digestibility of the fruit (unripe fruits contain less sugar and more difficult-to-digest complex carbohydrates than do ripe fruits). Thus, improving the discrimination of ripe fruit would have facilitated the digestive process, thereby decreasing the burden on the gut in frugivorous primates and increasing the energy available to support a larger brain. The gene duplication event which produced the third cone pigment in the ancestors of Old World monkeys, apes and humans may have favored the evolution of larger brains in this group. The capacity for color vision is also related to the functions of brain structures such as the parvocellular layers of the lateral geniculate nucleus, which relay to the visual cortex the opponent-color channels responsible for color discrimination. Recently Robert Barton found that the size of the parvocellular, but not the magnocellular, layers of the lateral geniculate nucleus is closely related to neocortex size.

The great expansion of the brain in humans has been accompanied by a commensurate reduction in digestive organ weight. There has been virtually a gram for gram trade-off between the expansion of the human brain and the reduction in the weight of our digestion organs relative to the apes. Carnivores, like frugivores, tend to have simple digestive systems because meat, like fruit, contains nutrients that are easy to digest. Thus the expansion of the brain in humans came with the benefit of an increased capacity to harvest rare, highly nutritious foods at the expense of the ability to consume common but less digestible materials.

## Brains and Time

Animals that have longer life spans are likely to experience more extreme environmental fluctuations and thus be exposed during their longer lives to more severe crises, such as shortages in normally used food resources, than are animals with shorter life spans. I was inspired to pursue this line of thinking by the picture of a very old capuchin monkey, Bobo, that appeared on the cover of *Science* magazine in 1982. Bobo was reported to be the oldest known monkey in captivity, and he ultimately lived to be nearly 54 years old. Bobo's great longevity intrigued me because capuchins have large brains and are famous for their ingenuity in the use of tools and their extraordinary capacity to capture the attention of humans by their clever antics. Capuchins are also long-lived in nature. John Robinson studied a population of about 200 capuchins in their native habitat in Venezuela for a 10-year period and found that their mortality rate was remarkably low relative to other monkey species. Capuchins, in fact, are the primates that lie closest to humans in the graph below, which plots relative brain weight and life span. I also remembered that a primatologist, Charles Janson, who had studied capuchins in the Peruvian Amazon, had told me that during periods of shortage, when the foods normally eaten by the capuchins became scarce, the older monkeys were able to harvest alternative foods because of their long experience in the forest. Since capuchins are extremely social, their harvesting expertise was readily communicated to other members of their troop. It was also clear to me that Bobo's longevity was not an isolated fluke because I knew of other capuchins who were still alive in their forties. By contrast, the longest surviving macaque monkeys, with relative brain sizes somewhat smaller than that of capuchins, were in their mid-thirties.

I decided to test the hypothesis that relative brain size is linked to longevity by obtaining life-span records for as many primate species as possible. This was an arduous process that took many

Bobo, a capuchin monkey, 47 years old at the time of this photograph, lived to the age of 54.

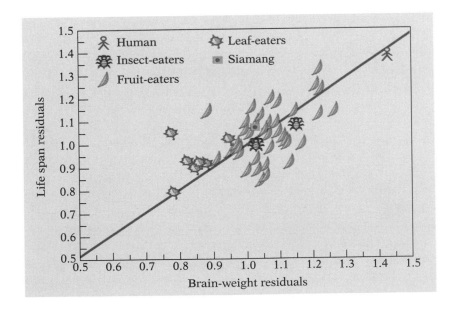

Life-span residuals versus brain-weight residuals for haplorhine primates (tarsiers, monkeys, apes, and humans). Based on an average of four reports, the siamang's diet consists of 43 percent fruit and 43 percent leaves.

## The Brain's Clock

The close relationship between brains and time indicates that the brain must have some sort of clock mechanism. The brain's "clock" is located in the suprachiasmatic nucleus, a small collection of neurons located in the hypothalamus just above the crossing of the optic nerves at the optic chiasm. Robert Moore showed that retinal fibers enter the suprachiasmatic nucleus and convey information about the amount of ambient light and therefore about the current stage in the day–night cycle. The nucleus contains pace-

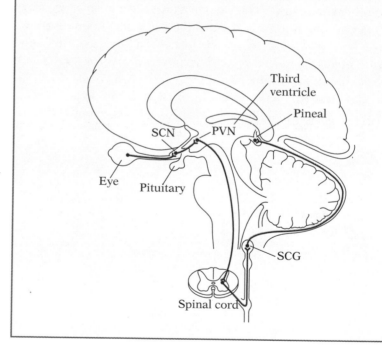

years of scanning and evaluating zoo records and the assistance of a large number of people, most notably my students Atiya Hakeem, Todd McLaughlin, and Gisela Sandoval, as well as Marvin Jones, who was the registrar for the San Diego Zoo. I decided to restrict the longevity search to primates because this group contains a manage-

maker neurons that determine the daily rhythm of the animal by regulating the secretion by the pineal gland of the hormone melatonin, which in turn controls the cycle of sleep and wakefulness. The pineal gland is homologous with the parietal eye in fish, amphibians, and reptiles.

The daily cycle is not exactly 24 hours and thus is termed a circadian rhythm, from the Latin words for "about a day." The circadian period varies in different animals. Michael Menaker and his colleagues have shown in studies in which they transplanted the suprachiasmatic nucleus from one animal to another that the transplanted nucleus imposes the donor animal's circadian rhythm on the recipient animal when its timing nucleus had been removed.

Opposite: Diagram of the human brain in midsagittal section showing the neural pathways by which photoperiodic information reaches the pineal gland, from the work of Robert Moore. Neurons in the retina project to the suprachiasmatic nucleus (SCN) of the hypothalamus, which in turn projects to the paraventricular nucleus, which projects to the spinal cord, which in turn relays to the superior cervical ganglion (SCG), which innervates the pineal gland.

The nucleus is also sensitive to annual changes in the day–night cycle and regulates seasonal changes in reproductive behavior. It also may have a role in controlling the total life cycle. Dick Swaab and his colleagues found that the number of neurons in the suprachiasmatic nucleus declines in the elderly and especially in patients afflicted with Alzheimer's disease. This loss of neurons disrupts the timing properties of the suprachiasmatic nucleus. The brain's clock runs down with time.

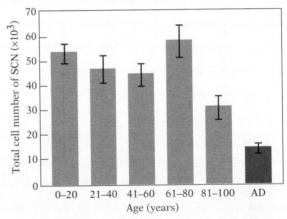

The total number of neurons in the superchiasmatic nucleus in different age groups and in sufferers from Alzheimer's disease (AD), from the work of Dick Swaab and his colleagues.

able number of species and a large variation in relative brain sizes. As a convenient measure of longevity, we looked at the age of the longest surviving individual of each primate species. Since longevity, like brain weight, is scaled to body weight, we calculated relative longevity in the same way as relative brain weight using the method

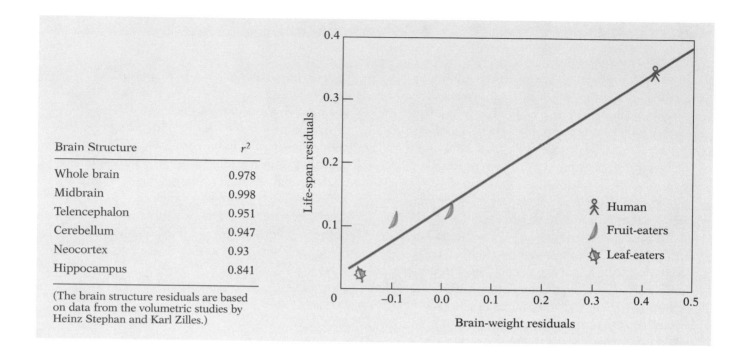

| Brain Structure | $r^2$ |
| --- | --- |
| Whole brain | 0.978 |
| Midbrain | 0.998 |
| Telencephalon | 0.951 |
| Cerebellum | 0.947 |
| Neocortex | 0.93 |
| Hippocampus | 0.841 |

(The brain structure residuals are based on data from the volumetric studies by Heinz Stephan and Karl Zilles.)

Life-span residuals versus brain-weight residuals for the gorilla (a leaf-eater), orangutan and chimpanzee (fruit-eaters), and human. The table shows the amount of the variance in the brain and life-span residuals that is accounted for by the relationship; $r^2$ is the correlation coefficient squared.

of residual variance. We found that relative longevity and brain weight are strongly related in monkeys, apes, and humans. The predominately leaf-eating primates are all clustered in the lower left part of the distribution and have smaller brains and shorter lives than the other primate species. The human data point lies close to the regression line, indicating that the maximum human life span is close to what would be expected for a primate of our relative brain size. The relative volumes of many brain structures, such as the neocortex, amygdala, hypothalamus, and cerebellum are also related to longevity. However, a number of brain structures are not related to life span; these are mainly the parts of the brain devoted in the early stages of sensory processing, such as the olfactory bulb, the lateral geniculate nucleus, and the vestibular nuclei. These structures are less involved in memory and strategies to cope with environmental variation than are the parts of the brain linked to longevity.

When we considered the data for humans and our closest relatives (gorillas, orangutans, and chimpanzees) we found that the relationship between relative brain size and longevity was very strong. This was also true for nearly all the brain structures for which we have volumetric data for these primates. The same dietary skewing is present in this smaller sample as in the larger one. The gorilla, a specialized leaf- and plant-eater, has a relatively smaller brain and shorter life span than do the other primates. Within primates, the strength of the correlation between brain and longevity increases with phylogenetic affinity to humans. It is weak among the prosimians, stronger among the monkeys, and very strong in the group comprised of the great apes and humans.

## The Social-Brain Hypothesis

In *The Descent of Man,* Darwin proposed that the evolution of intelligence is linked to living in social groups. In 1982, I proposed a variant of the "social-brain" hypothesis emphasizing the role of brain structures involved in social communication. In 1988, Richard Byrne and Andrew Whiten proposed another version of the social-brain hypothesis in a book entitled *Machiavellian Intelligence* in which they proposed that the development of social expertise was a key factor in the evolution of the brain in primates. The social-brain theory has sometimes been referred to as the "Machiavellian" hypothesis, with the implication that it requires the craft and cunning of a Renaissance prince to cope with the complexities of primate social life. The ideal data to test the social-brain hypothesis would be a measure of the complexity of social interactions in different species. Unfortunately, a direct measure does not exist to compare different species, but there is good data on social-group size for many species, which is rough measure of the complexity of social life. Atiya Hakeem, Andrea Hasenstaub, and I tested the social-brain hypothesis by examing the relationships between group sizes and the relative sizes of the brain and its component structures in different species. To measure the relative brain size, we used the same residuals method that we used to test the relationships between brain and life span. We found no significant relationship between group size and the weight of the brain or the volumes of its components. We were particularly surprised to find no relationship between group size and the amygdala because of the strong evidence for the amygdala's role in social communication. Recently, Robin Dunbar has also tested the social-brain hypothesis, and he too found no relationship between most measures of brain size and social-group size, but with one exception.

A nineteenth-century wood engraving of an adult chimpanzee.

He found a significant relationship between group size and the ratio of neocortex volume to the volume of the rest of the brain. The larger the size of the neocortex relative to the rest of the brain, the larger the social-group size in primates. We have confirmed his finding with a larger sample that included all the primate species for which there were neuroanatomical data. We found that social-group size predicted about 45 percent of the variation in the neocortex ratio among primate species. We performed the same analysis using the life-span data and found that it predicted about 47 percent of the variation in the neocortex ratio. Group size was related to less than 3 percent of the variation in life span so that these two measures are essentially independent of each other. We have no clear explanation as to why social-group size is related to one measure of relative neocortex size but not to the others. Longevity has a much wider and more consistent set of relationships with the brain and its component structures than social-group size has.

The evolution of social complexity might be linked to other neurobiological changes apart from brain size. For example, Klaus-Peter Lesch found a class of promoters regulating the expression of the serotonin transporter that is unique to anthropoid primates. Variations in this class of promoters are linked to variations in personality traits that may be advantageous in the complex anthropoid societies.

## Big Brains and Parenting

Having a larger brain is linked to enhanced survival. This being the case, why don't more animals have large brains? The answer to this puzzle is that the costs of growing and maintaining a big brain are very high both for the individual and for its parents. In a newborn human the brain absorbs nearly two thirds of all the metabolic energy used by the entire body. This enormous burden results from the very large relative size of the brain in human infants and from the additional energy required for dendritic growth, synapse formation, and myelination, which is far greater even than the considerable energy required to maintain the adult brain. Because the brain requires nearly two thirds of the infant's energy supply, this constraint probably sets an upper limit in the evolution of brain size because the muscles and the other vital organs, the heart, the liver, the kidneys, and the digestive organs, must use energy as well.

Nurturing a large-brained baby imposes enormous energy costs on the mother because of the burden of lactation, which is far more costly than gestation. In small mammals lactation can triple the mother's food requirements. The nutritional constituents of breast

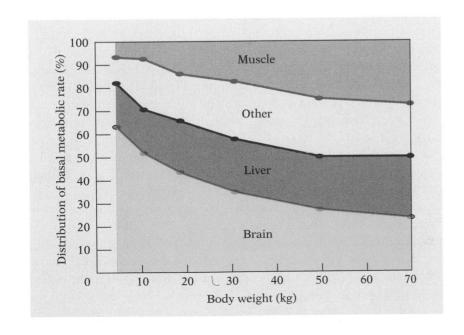

Distribution of brain, liver, and muscle metabolic rate as percentages of total basal metabolic rate at different body weights in humans, from the work of M. A. Holliday.

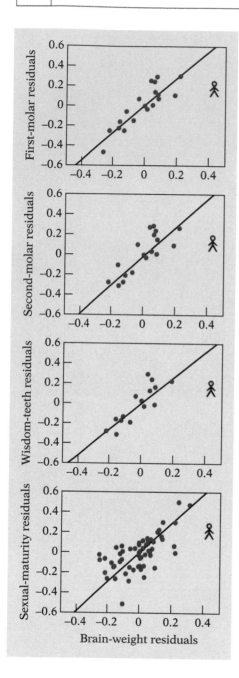

Relationships between measures of maturation time and brain-weight residuals in different primate species. The ages at which different types of teeth erupt are good markers for developmental stages. Note that humans consistently mature much earlier by each of these measures than would be expected on the basis of relative brain size. The slopes of the regression lines are nearly identical, indicating a consistent relationship between relative brain size and development time. The teeth data were collected by Holly Smith and her colleagues; the average age of sexual maturity is taken from the compilation by Noel Rowe.

milk are probably optimized for brain growth in particular species. In a carefully controlled study of children tested at age 8, those who had been bottle-fed human milk as babies had an average IQ 10 points higher than did the children who had been fed formula.

Not only are the energetic costs high, but development is slow in big-brained babies. George Sacher proposed that the brain serves as a pacemaker for the growth of embryos. In primate species, relative brain mass scales with the time after birth required to reach maturity, implying that the development of larger brains requires more time.

The additional time is needed for the postnatal growth of the brain, which in humans reaches its full adult size only by about the time of puberty. This postnatal growth includes the formation of myelin insulation around axons, which proceeds at different rates in different parts of the brain. Paul Flechsig showed that the axons of subcortical structures acquire their myelin insulation before those of the cortex, and within the cortex the primary sensory areas are myelinated long before the higher cortical areas in the temporal, parietal, and frontal lobes. The process of myelin formation is very slow. In magnetic resonance imaging studies, Tomas Paus and his colleagues found that parts of the forebrain continue to myelinate until age 17.

The rate of synapse formation also varies among cortical areas. Peter Huttenlocher found that synaptogenesis is much slower in the frontal cortex than in the primary visual cortex. Time is also required for the formation of experience-dependent connections essential for adult functioning. For example, as discussed in Chapter 6, the capacity to judge the size and distance of objects develops very slowly and is still quite immature in 8-year-old children. The gradual refinement of this capacity probably depends on countless interactions between

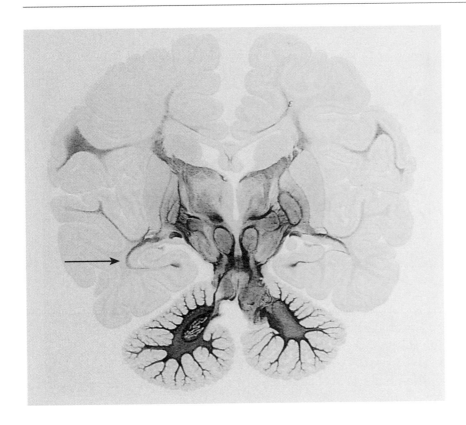

The myelinating pathways in a 7-week-old human infant, from the work of Paul Flechsig. This is a horizontal section through the forebrain and cerebellum; myelin is stained blue. Note that the myelinated pathways are already well developed in the cerebellum and the central parts of the brain at this stage, but there is relatively little myelin in the white matter associated with the neocortex. However, there is a U-shaped pathway (arrow) of myelinating fibers leading from the lateral geniculate nucleus of the thalamus to the primary visual cortex. Bands of fibers also lead to the primary somatosensory and motor cortical areas. The fiber connections of the higher cortical areas myelinate much later in development.

the child and his or her spatial environment, which in turn influences synaptic changes in the visual cortex that continue quite late in childhood. Because the brain is unique among the organs of the body in requiring a great deal of feedback from experience to develop to its full capacities, brain maturation may serve as a rate-limiting factor that governs the maturation of the entire body As Steven Quartz and Terrence Sejnowski have suggested, the animal's experience in interacting with its environment directs the growth of dendrites and the formation of synaptic connections. They propose that learning is a process that occurs in successive stages, each building on the earlier ones. Larger brains require a longer time to develop because more stages are involved.

Thus the rearing of large-brained babies requires parental support for commensurately long periods. Moreover, large-brained offspring are mostly single births and the interbirth intervals are long, which probably reflect the large costs of rearing these offspring. The parents must live long enough past their sexual maturity to sustain

A young orangutan, painted by Richard Owen.

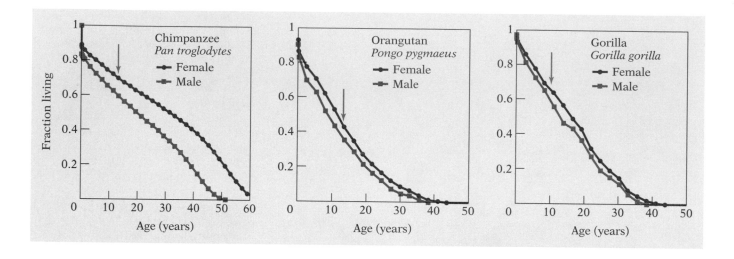

Differential survival between male and female apes. The chimpanzee data are from the work of Bennett Dyke and his colleagues; the orangutan and gorilla data were compiled from zoo records by Roshan Kumar, Aaron Rosin, Andrea Hasenstaub, and the author. The arrow indicates the average age at which females give birth to their first offspring. The graphs show that at every age there are fewer surviving males than females.

the serial production and maintenance of a sufficient number of offspring to replace themselves while allowing for the early death or infertility of their children. Therefore, I hypothesized that in large-brained species that have single births the sex that bears the greater burden in the nurturing of offspring will tend to survive longer. If the caretaking parent dies, the orphan has a high probability of dying, but if the noncaretaking parent dies, this event will have little impact on the offspring's chances of survival. The death of a noncaretaking parent may even enhance the survival of its offspring by removing

a competitor for scarce food resources. Thus genes enhancing the survival of the caretaking parent will be favored by natural selection, since they will be more likely to be transmitted to the next generation than genes that might enhance the survival of the noncaretaking parent. Male primates are incapable of gestating infants and lactating; but in several species, fathers carry their offspring for long periods, and the young may stay close to the father even after they move independently. According to the caretaking theory, females should live longer than males in the species where the mother does most or all of the care of offspring; there should be no difference in survival between the sexes in species in which both parents participate about equally in infant care, and in those few species where the father does a greater amount of care than the mother, males should live longer. Roshan Kumar, Aaron Rosin, Andrea Hasenstaub, and I tested this hypothesis by constructing mortality tables similar to those used by the life insurance industry for male and female anthropoids (monkeys, apes, and humans) and comparing these data with the sexual division of care for offspring.

The great apes are our closest relatives. Chimpanzees, orangutans, and gorillas nearly always give birth to a single offspring and the interval between births ranges from 4 to 8 years. Female chimpanzees, orangutans, and gorillas have a large survival advantage in data obtained from captive populations.

For example, in captivity the average female chimpanzee lives 42 percent longer than the average male. In the case of chimpanzees there also are data available from populations living in nature. In a 22-year study of a population of 228 chimpanzees living in the Mahale Mountains near the shores of Lake Tanganyika, Toshisada

An orangutan mother with her offspring. Except for mothers and their offspring, orangutans lead a solitary existence. The burden of taking care of the slowly maturing offspring falls entirely on the mother. Birute Galdikas found that the average interbirth interval for orangutan mothers is 8 years.

Opposite: A chimpanzee family studied by Jane Goodall at Gombe. The mother, Flo, was about 40 years old when this photograph was taken. Her infant, Flint, snuggles securely in her arms. Flo's adult daughter, Fifi, looks on while the adolescent Figan grooms his mother. When Flo died a few years later, Flint, then 8 years old, died shortly thereafter, apparently unable to survive without her support. Goodall (1986) found that about half of orphaned chimpanzees died and that the surviving orphans often exhibited retarded sexual maturation. Maternal survival also enhances the success of adult offspring. In her study at Gombe, Goodall noted that Flo's forceful personality contributed to the high status of her adult offspring. Similar maternal contributions to the success of adult offpsring have also been observed in long-term studies by Toshisada Nishida for the chimpanzees at Mahale and by Takayoshi Kano for bonobos at Wamba. Male chimpanzees rarely care for their offspring. These factors would lead to natural selection favoring genes that would enhance female survival.

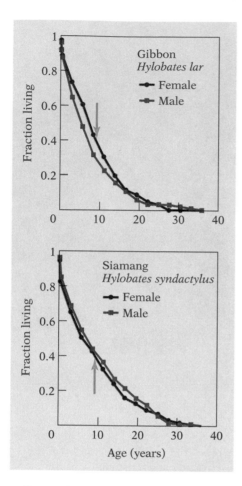

Differential survival patterns in gibbons and siamangs, closely related species living in the same habitat. Note that the female gibbons outlive males, but that male siamangs slightly outlive females. Siamang fathers are the only apes that carry their offspring on a regular basis. The data were compiled from zoo records by Roshan Kumar, Aaron Rosin, Andrea Hasenstaub, and the author.

Nishida and his colleagues found an equivalent number of male and female births but three times as many females as males in the adult population. This difference was not due to differential patterns of migration, and thus their observations indicate a strong female survival advantage for chimpanzees living in the wild. Chimpanzee mothers generally provide nearly all the care for their offspring, and females possess a very strong survival advantage. Although male care of infants is rare in chimpanzees, Pascal Gagneux and his colleagues have observed instances in which males have adopted orphaned infants and cared for them. Their observations indicate that the potential for male care is present in chimpanzees though rarely expressed. Orangutan mothers provide all the care for their offspring, which have very little contact with the solitary adult males. Gorilla mothers provide most of the care for their offspring, but the fathers protect and play with them. The female survival advantage in gorillas, while significant, is not so large as in chimpanzees or orangutans.

The lesser apes are our next closest relatives. Gibbons and siamangs live in pairs and have a single baby about once every 3 years. They maintain their pair bonds and defend their territories through spectacular vocalizations similar to the pair-bonding songs of birds. Gibbon mothers provide nearly all the care for their offspring, but David Chivers found that siamang males play a much larger parental role than do gibbon males. Siamang mothers carry their infants for the first year, but during the second year males carry the growing infant. Siamang males are unique among apes in carrying their infants and in the closeness of their bonding with their offspring. Gibbon females have a survival advantage over males, but the situation is reversed in siamangs, where the males have a small advantage. Gibbon females on average live about 20 percent longer than males, but siamang males live 9 percent longer than females. Siamang fathers are the only male apes that carry their infants and the only apes in which males outlive females.

In Old World monkeys, females do most of the infant care, and several studies from natural populations show a female survival advantage. In New World monkeys, we found a significant survival advantage in captive spider monkeys, and John Robinson found a female survival advantage in the natural population of capuchin monkeys observed in Venezuela. In both spider and capuchin monkeys, mothers do virtually all the infant care. However, the situation is dramatically reversed in two other New World primates, the owl

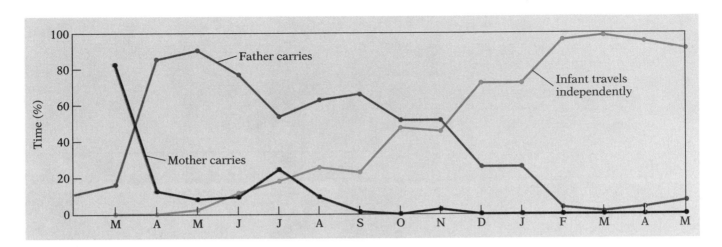

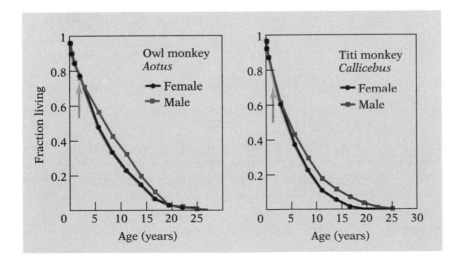

monkeys and titi monkeys. These monkeys live in pairs, like gibbons and siamangs, and also maintain their pair bonds and defend their territory through vocalizations. The fathers carry their infants from shortly after birth except for brief nursing periods on the mother and occasional rides on older siblings. I have observed in my colony of owl monkeys that if the father dies, the mother will not carry the infant, and thus the survival of the infant depends on the father. In both owl and titi monkeys, males and females die at the same rate until maturity, but after maturity the males have a survival advantage over females. Thus the timing of the male survival

The percentage of time that a siamang father carries his offspring during its second year, from observations by David Chivers.

The adult male survival advantage in owl monkeys and titi monkeys, species in which the fathers carry their infants from shortly after their birth. The data were compiled from zoo records by Roshan Kumar, Aaron Rosin, Andrea Hasenstaub, and the author.

An owl monkey father with his infant. In her studies in the Peruvian Amazon, Patricia Wright found that owl monkey and titi monkey fathers carry their infants from almost immediately after birth and provide most of their care. The fathers also share food with their offspring. By sharing food, the father shows his offspring which foods are nontoxic, and thus he provides not only nutrition but also information vital for the survival of his offspring. By contrast, the mothers tend to lead family progressions through the forest, and thus they expose themselves to greater risks than the fathers. Although owl monkeys and titi monkeys are similar in size, monogamy, and parenting behavior, DNA sequence data, analyzed by Horacio Schneider and Morris Goodman and their colleagues, indicate that they are not closely related. Thus extensive paternal care with enhanced male survival is a specialization that probably evolved independently in lines leading to modern owl monkeys and titi monkeys.

advantage corresponds to the period in their lives when they carry their offspring.

It is well known that women tend to live longer than men. It is often assumed that this is a modern phenomenon resulting from the greatly reduced risk of death in childbirth and other improvements in women's health practices. However, the female survival advantage is present in the oldest systematic records from a human population, which were collected in Sweden beginning in 1780, long before modern health practices were instituted. The female

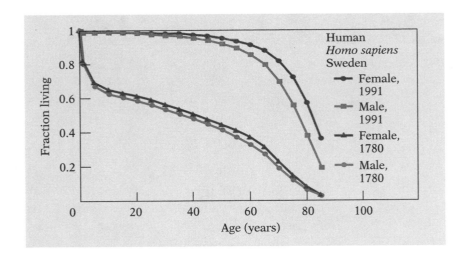

The human female survival advantage in the Swedish population in 1780 and 1991, plotted from data in the demographic study by Nathan Keyfitz and Wilhelm Fleiger and from the United Nations demographic database. According to current World Health Organization data, females live longer than males in all but 2 of 171 reporting nations.

advantage is present at every age and for every Swedish census since 1780. In the Swedish population women live 5 to 8 percent longer than men. Similar female advantages were recorded in the earliest data from England and France in the nineteenth century and a female advantage has been present in most nations throughout the world in the twentieth century. A female survival advantage has also been found for adults in the Aché, a well-studied hunter-gather population living in the forests of eastern Paraguay. These data strongly suggest that the survival advantage in human females has deep biological roots. However, it is smaller in relative terms than in gorillas, gibbons, orangutans, spider monkeys, and chimpanzees.

In most species there is a female advantage throughout life, but in all the anthropoids in which there are single births and the males carry their offspring, there is either no difference in survival between the sexes or there is a definite male survival advantage. These results run counter to the reasonable expectation that taking care of an infant would decrease rather than increase chances for survival. The magnitude of the difference in survival corresponds to the difference in the amount of care given to the offspring by each sex. Thus in the great apes where the mothers do virtually all the care, there is a large female advantage. Human males contribute significantly, but human females are the primary caregivers, and in humans there is a proportionally smaller, but still sizable, female advantage. In Goeldi's monkeys both sexes provide about the same amount of care and

| PRIMATE | FEMALE/MALE SURVIVAL RATIO | MALE CARE | | |
|---------|---------------------------|-----------|---|---|
| Chimpanzee | 1.418 | Rare | | |
| Spider monkey | 1.272 | Rare | | |
| Orangutan | 1.203 | None | | |
| Gibbon | 1.199 | Pair-living, but little direct role | | |
| Gorilla | 1.125 | Protects, plays with offspring | | |
| Human (Sweden, 1780–1991) | 1.052–1.082 | Supports economically, some care | Increasing male survival ↓ | Increasing male care ↓ |
| Goeldi's monkey | 0.974 | Both parents carry infant | | |
| Siamang | 0.915 | Carries infant in second year | | |
| Owl monkey | 0.869 | Carries infant from birth | | |
| Titi monkey | 0.828 | Carries infant from birth | | |

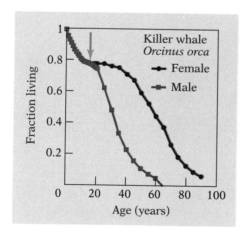

Female killer whales live much longer than males. This graph was plotted from demographic data for a natural population of killer whales living in Puget Sound collected by Peter Olesiuk and his colleagues.

there is no difference in survival. In siamangs, both parents participate with the father taking over in the later stages of infant development, and siamang males have a small advantage. In owl monkeys and titi monkeys, males carry the babies most of the time from shortly after birth, and thus infant survival depends substantially on the male; in these monkeys there is a large male advantage.

Similar data have come from a nonprimate, big-brained species. Killer whales have very large brains. Their calves are born singly with an interbirth interval of 5 years, and they remain in close association with the mother throughout their lives. Males appear to have little direct role in parenting. A long-term demographic study of a natural population of killer whales in Puget Sound found that female life expectancy is more than 20 years longer than in males. The average female lives about 75 percent longer than the average male.

The differential mortality between caretakers and noncaretakers may be in part because the former are risk-averse and the latter tend to be risk-seeking. Caretakers tend to avoid risk because they risk not only themselves but also their offspring. This may be a conscious decision or the result of genetically determined instincts that would be favored by natural selection because they would lead to more surviving offspring. A second major factor may be a differential vulnerability to the damaging effects of stress. Natural selection would also favor the evolution of genes in caretakers that protect them against the damage induced by stress. The ratio between the rates at which males and females die varies during the course of life. In humans, the female survival advantage begins shortly after conception and continues throughout life, with the largest advantage, in terms of the size of the ratio between male and female age-specific death rates, occurring at around age 25. In many countries, a second peak in the male-to-female ratios appears later in life. This second peak is present in the Swedish census data, extending back to 1780. The two peaks also are present at about the same stages in the life cycle in some nonhuman primates such as gorillas and gibbons. The peak in early adulthood corresponds approximately to the period of greatest responsibility for child care in women. The second peak appears to be related to a higher risk of heart disease and other afflictions in men. I believe that these two peaks represent two underlying mechanisms, one of which is mainly acting on the young and the other on the old. The first peak is largely due to differences between males and females in risk-taking behavior that results in higher rates resulting from accidents and violence in younger males. The second peak may result from increased male vulnerability to pathological conditions that develop without overt symptoms over a long period of time, such as genetic damage ultimately leading to cancer, as well as high blood pressure and clogged arteries, all of which may be related to the cumulative effects of stress. By contrast, in owl monkeys and titi monkeys, the male survival advantage emerges shortly after maturity, at the time when fathers begin to care for their offspring. This hypothesis would predict that their enhanced survival may be due to reduced risk-taking and vulnerability to stress.

In the contemporary United States population, females have lower risks than males of dying from the 13 most prevalent causes of death, indicating that the female survival advantage has an extremely broad base. A hormonal basis for this effect is evidenced by the observation by Francine Grodstein and her collaborators that post-menopausal women who currently receive estrogen replacement have a lower risk of death as compared with post-menopausal women who have never received supplemental estrogen. Estrogen enhances the actions of serotonin and thus may be responsible for reducing risk-taking behavior, as discussed in Chapter 2. Melanie Pecins-Thompson and her colleagues found in macaque monkeys

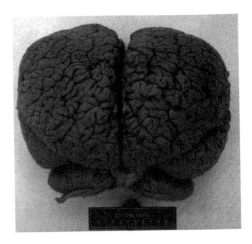

The enormous brain of a killer whale. This brain, which weighs 7100 grams, came from a whale weighing 4455 kilograms. Dissection and photograph by Raymond Tarpley and Sam Ridgway.

A young killer whale with its mother. Killer whales take about 15 years to mature; the average interbirth interval for killer whale mothers is 5 years.

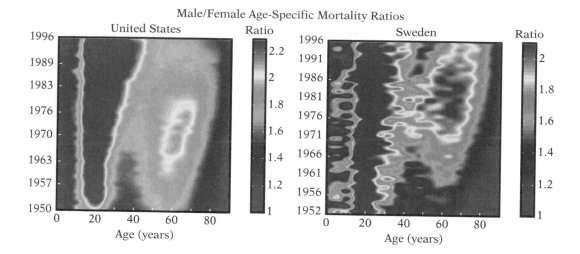

Male/Female Age-Specific Mortality Ratios

Excess male deaths as a function of age from 1950 to 1996 in the United States and Sweden. Similar patterns are present in the data for Japan, Canada, and many other countries with well-developed health-care systems. The red pattern in the young adult years indicates that more than twice as many men as women die at this stage of life. The pattern is much smoother for the United States because of the much larger population size. The earlier Swedish data, going back to 1780, consistently show similar peaks in early and late adulthood although the peaks are not as large as for modern data. This consistency suggests that biological factors are partially responsible. The second peak occurs after child rearing but reflects differential responses to stress earlier in life. The analysis was done by Andrea Hasenstaub and the author; the data are from the demographic study by Nathan Keyfitz and Wilhelm Fleiger and the United Nations demographic database.

that estrogen inhibits the expression of the gene that makes the transporter protein responsible for serotonin reuptake. Thus estrogen acts like drugs such as Prozac, which inhibit the removal of serotonin from the synaptic cleft and consequently increase the synaptic concentration of serotonin. Because of estrogen's effects on the serotonergic system, it has been called nature's psychoprotectant.

Another possible basis for differential survival may be related to the stress hormones, the corticosteroids. The clearest evidence for this comes from the work of Robert Sapolsky, who encountered and studied a group of vervets that had previously been subjected to chronic stress caused by crowded living conditions. Vervets are a type of monkey in which females do most of the care for offspring. Sapolsky found a substantial loss of neurons in part of the cerebral cortex, the hippocampus, in males but not in females. The hippocampal neurons are richly supplied with receptors for the corticosteroid hormones, which are produced by the adrenal cortex to mobilize the body's defenses when subjected to stress. One role of the hippocampus is to regulate the pituitary's secretion of adrenocorticotropic hormone, which in turn signals the adrenal cortex to secrete the corticosteriod hormones into the bloodstream. The secretion of the corticosteroid hormones is the body's way of responding to sudden, life-threatening emergencies, but the chronic secretion of these hormones can be very damaging. The hippocampal neurons are

| Leading Causes of Death in the United States in 1995 | | |
|---|---|---|
| RANK | CAUSE | MALE/FEMALE MORTALITY RATIO |
|  | All causes | 1.7 |
| 1 | Heart disease | 1.8 |
| 2 | Cancer | 1.4 |
| 3 | Stroke | 1.2 |
| 4 | Emphysema and bronchitis | 1.5 |
| 5 | Accidents | 2.5 |
|  | Car | 2.3 |
|  | Other | 2.9 |
| 6 | Pneumonia and influenza | 1.6 |
| 7 | Diabetes | 1.2 |
| 8 | HIV/AIDS | 5.0 |
| 9 | Suicide | 4.5 |
| 10 | Liver disease and cirrhosis | 2.4 |
| 11 | Kidney disease | 1.5 |
| 12 | Homicide | 3.7 |
| 13 | Septicemia | 1.2 |
| 14 | Alzheimer's disease | 1.0 |

(Data from *Monthly Vital Statistics Report*, 1997.)

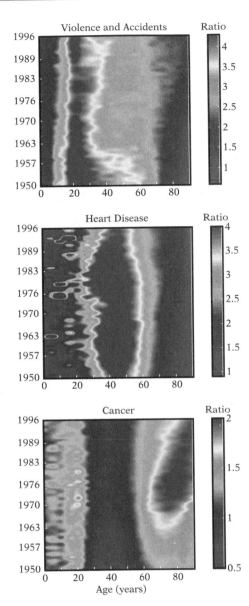

Ratio of male to female deaths as a function of age and calendar year in the United States. Excess male mortality from violence and accidents peaks around age 25, from heart disease at 50, and from cancer at 75. The first peak is related to risk-taking; the heart disease and cancer peaks may reflect male susceptibility to stress.

particularly vulnerable because they have many receptors for these hormones. In other systems corticosteroids activate the caspases, the cellular self-destruct mechanism, and conceivably they also trigger this mechanism in hippocampal neurons. Corticosteroids also suppress serotonin receptors in hippocampal neurons, which may diminish their stability and further increase their vulnerability. Because the serotonin re-uptake mechanism is inhibited by estrogen, males may be more vulnerable than females in some species. The loss of the hippocampal neurons due to hyperexcitation means that the brakes on the secretion of the stress hormones are burned out, leading to escalating levels of damage and ultimately to death. Sapolsky's results indicate that male vervets are much more vulnerable to the destruction of the brain's system for regulating the stress

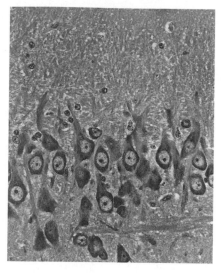

Normal

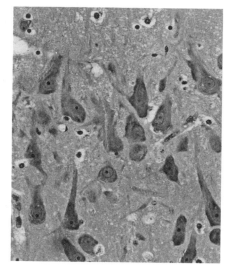

Stressed

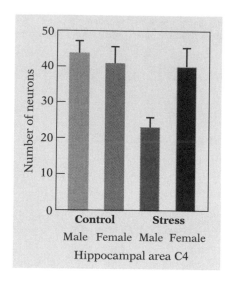

Neuron loss in the hippocampus of stressed male monkeys. The left photomicrograph is from the hippocampus of a control monkey; the right photomicrograph, from the same place in the hippocampus of a stressed male, shows a loss of neurons and dendritic atrophy in the remaining neurons. The graph shows the number of neurons in samples of hippocampal area CA4 in unstressed male and female controls and in stressed males and females. Robert Sapolsky and his colleagues also found similar neuronal losses in the other CA fields of the hippocampus of stressed males. In these monkeys, the stress resulted when they were captured by the Kenyan government at the request of farmers and housed under crowded conditions.

response than are females. This may be the mechanism for male vulnerability in other species where females are the primary caregivers, and this theory predicts that the opposite would be true for those species where males are the primary caregivers.

What is the biological role for the higher level of risk-taking in males in some species? In *The Descent of Man*, in a section entitled the "Law of Battle," Darwin linked male aggression to competition among males for females. This has led to the widely accepted idea that aggressive males become socially dominant and because of their dominance enjoy greater sexual access to females and therefore greater reproductive success. However, there is evidence to indicate that other factors may be involved in male risk-taking.

Let us begin by examining the first part of this relationship: does aggression lead to social dominance? In Chapter 2, I discussed the changes in social status in male vervet monkeys induced by experimentally manipulating serotonin levels. In this study, male status was invariably preceded by changes in affiliative behaviors with females in the social group such as grooming interactions. Increased affiliative behavior led to increased female support in dominance interactions with other males, which in turn led to rising status. Decreased affiliative behavior led to decreased female support, which in turn led to declining status. This investigation and many observational studies indicate that high status in primate groups is much more dependent on social skills and coalition building than on aggression.

Now let us turn to the second part of the aggression–dominance–reproductive success theory: does the possession of high rank lead to reproductive success? Pascal Gagneux and his colleagues have conducted a long-term study of the social structure of chimpanzees living in the Tai forest in the Ivory Coast. In order to measure male lineages, they extracted DNA from hair samples for all the members of this group, and thus they were able to determine which chimpanzees had fathered which offspring. They found two surprising results. First, on the basis of the DNA patterns they were able to rule out all the males in the group as possible fathers of half of the youngsters. Thus the females were covertly mating with males *outside* their social group; the status of those males within their own groups is unknown. Second, for the youngsters that were fathered by males within the social group, there was only a weak relationship between dominance and reproductive success. Brutus, the top-ranking male for 10 years, and Macho, who was the alpha male for 1.5 years, sired no offspring during their periods of dominance, although each sired one after they declined in status. These results highlight the importance of actually determining male parentage through DNA studies, because it is only through such studies that male reproductive success can be determined, which is crucial for measuring the influences of different behaviors on the evolutionary process. Until there is a substantial body of genetically established data for a number of carefully observed primate species, the role of male dominance in reproductive success will remain undetermined. However, observations by Sapolsky in baboons does suggest that high male status does confer a different advantage. He found that the levels of cortisol, a corticosteroid hormone, are inversely related to social status. Therefore, high-status males are less at risk to adverse consequences of this hormone. Important advantages of high status in males are reduced vulnerability to the deleterious effects of stress and better access to food resources.

There is strong evidence that high status does confer reproductive success in female chimpanzees, and it is clear that social competence plays an important role in determining the female dominance hierarchy. Goodall and her collaborators found that the offspring of high-status females are more likely to survive and that they mature at an earlier age. They also found evidence that the

Darwin proposed that large male body size is related to male competition, and therefore the ratio of male to female body weight has been seen as an indicator of the intensity of male aggression and competition. Andrea Hasenstaub and I examined the possibility that male/female weight ratio might be inversely related to male/female survival ratio under the premise that fighting might account for some of the excess male mortality. We found a slight tendency for the survival ratio to be related to the weight ratio, but the effect was not statistically significant. Moreover, this tendency is not seen in the two most closely related primates in the table on page 184, gibbons and siamangs; it is the siamangs, with the higher weight ratio, that exhibit male care and a higher male/female survival ratio. In addition, spider monkey males and females are the same size, but females live 27 percent longer than males. Therefore excess male body size does not account for excess male mortality.

## Cooperative Male Care in Marmosets and Tamarins

Marmosets and tamarins, which are small New World monkeys, have many more offspring than do other monkeys and have an unusual solution for providing care for their infants. Unlike other monkeys, which have single births, marmosets and tamarins usually give birth to twins or sometimes triplets. Shortly after birth, females become sexually receptive and can conceive again. Thus marmosets and tamarin females can produce up to six babies per year. These primates have developed a different way to nurture their multiple, slowly developing, large-brained infants. Marmosets and tamarins live in extended families in which everyone and especially the males participate in infant care. Marc Van Roosmalen has even observed a male assisting in the birth process by cutting the umbilical cords and eating the afterbirth. Paul Garber found that the presence of up to four males in the family enhances the survival of the infants.

The males cooperate in caring for the infants in their group, and there is little aggression among males within the family. The males are very strongly attracted to the infants; they carry them whether or not they are their biological offspring, and they share food with

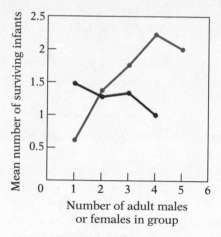

Surviving infants based on number of adult males

Surviving infants based on number of adult females

The graph shows that infant survival in tamarins increases as a function of the number of caretaking males in the extended family groups; having more females results in a slight reduction in the number of surviving infants. This graph, from the work of Paul Garber, is based on observations of 47 extended tamarin families living in nature.

high-status females live longer than the low-status females. These effects may be the consequence of less stress and better access to food and other resources in the high-status females.

Social competence probably counts for more than aggression in achieving either high status or reproductive success in primates. Why then are the noncaretaking males aggressive and prone to risk-taking? Why would natural selection favor the evolution of behaviors that increase the risk of dying? I think the answer is that risk-takers

them. I have even observed a male kidnapping the offspring of another family so as to carry it. Because of the cooperative care, offspring are less dependent on the survival of a particular caretaker. In our studies thus far we have found little difference in the survival of male and female marmosets and tamarins. Through intensely cooperative care of infants, marmosets and tamarins have overcome the evolutionary constraints imposed by large brain size.

A extended marmoset family enjoying a quiet moment.

constantly probe their world, seeking out new opportunities and detecting hazards in a constantly changing environment. Through their probing they generate new information that they communicate to close kin, thus enhancing their kin's survival and the propagation of their shared genes. Specific vocalizations for types of food and types of predators serve this communicative function. The risk-takers may also be crucial to colonizing new habitats during changing environmental conditions.

Both the evolution of large brains and the evolution of temperature homeostasis, as discussed in Chapter 5, required new developments in parenting behavior. Warm-blooded infants are dependent and cannot grow without parents to provide warmth and nutrition. Increasing brain size slows down post-natal development as measured by the ages at which different teeth erupt and by the age of sexual maturation. Large-brained, slowly developing, dependent offspring require long-surviving parents to reach maturity. A measure of this parental dependency effect is the differential survival of caretakers versus noncaretakers. In primates, the caretaker effect has a large influence on the patterns of survival with as much as a 42 percent female advantage when males have little role in nurturing offspring versus as much as a 20 percent male advantage when males carry offspring from soon after birth. The male caretaking effect is not as large because only females provide nutrition for their slowly developing offspring through lactation. The mechanisms responsible for the survival differences between caretakers and noncaretakers may ultimately be related to neurochemical differences that favor risk-aversive behavior in caretakers and risk-seeking behavior in noncaretakers, as well as greater vulnerability to the damaging effects of stress in noncaretakers.

## Brain Evolution in Hominids

The same ancestral stock of apes that gave rise to orangutans, gorillas, and chimpanzees also produced the hominids, the group that includes our immediate progenitors. The earliest hominids emerged about 4 million years ago in East Africa. Their brains were the same size, about 400 grams, as their ape cousins', but they walked bipedally, although perhaps having more capacity for tree climbing than do modern humans. We know that these early hominids, the australopithecines ("southern apes"), were bipedal because of the similarity of their pelvis and leg bones to our own and from a remarkable series of footprints which they left behind 3.7 million years ago in freshly fallen volcanic ash at Laetoli in modern Tanzania. A large number of australopithecine remains have been recovered, and from them it is possible to deduce that the males weighed about 40 kilograms while female weighed only about 30 kilograms. Thus males were about one third larger than females. A considerably larger male body size is also characteristic of the living great apes and was probably present in the last common ancestors of the great apes and hominids. For example, the average male chimpanzee is 35 percent heavier than the average female. In living mammals that have polygynous mating systems, males are considerably larger than females.

There is strong evidence for a worldwide shift to a cooler and drier climate about 2.3 million years ago. Steven Stanley has proposed that this alteration resulted from the formation of the land bridge between North and South America at the Isthmus of Panama, which changed the global pattern of ocean currents. This was a time of extensive glaciation and dust deposition in many parts of the world. In Africa, forests shrank and savanna-adapted species of antelope and small mammals replaced forest-adapted species. At about this time the early australopithecines gave rise to the robust australopithecines, which had massive jaws and molar teeth, and to the first humans. Body size remained about the same as in the earlier australopithincines, but brain size increased in both descendant groups. *Australopithecus robustus* had a 500-gram brain and the earliest humans, *Homo habilis* ("handy man"), had about a 600-gram brain. The *Homo habilis* remains from Olduvai Gorge show a smaller difference between male and female body size than in the australopithecines. Henry McHenry estimated that the Olduvai *Homo habilis* males weighed 37 kilograms and the female weighed 31.5 kilograms. Thus the males were about 17 percent heavier than the females, which is close to the ratio in modern humans. Shortly afterward, *Homo habilis* gave rise to *Homo erectus*, whose body size was close to that of modern humans but whose brain weighed about 800 to 900 grams, which is about two thirds of the brain size of modern humans. The size difference between males and females in *Homo erectus* was about the same as in modern humans, and their bipedal locomotion was evidently very efficient, because soon after their appearance in East Africa their descendants had migrated as far as East Asia, where their 1.8-million-year-old remains have been discovered. These early humans also fashioned primitive stone tools; it is possible that *Australopithecus* also made tools. The use of stone tools has been observed extensively in chimpanzees by W. C. McGrew and others. Tool manufacture does not appear to be so uniquely associated with human status as was once supposed.

The living great apes are found only in rain-forest habitats, the most stable environments on the planet. Australopithecines generally occupied woodland habitats. Early human remains are found in a more diverse range of habitats than are those of other hominids. In very dry habitats, there were no trees to provide escape from predators, and access to water was a special problem. Thus early humans

| Hominid Changes in Body Weight and Brain Weight | | | |
|---|---|---|---|
| SAMPLE | TEMPORAL RANGE (YEARS AGO) | BODY WEIGHT (KG) | BRAIN WEIGHT (G) |
| Living worldwide | — | 58.2 | 1302 |
| Late Upper Paleolithic | 10,000 – 21,000 | 62.9 | 1412 |
| Early Upper Paleolithic | 21,000 – 35,000 | 66.6 | 1460 |
| Late archaic *Homo sapiens* | 36,000 – 75,000 | 76.0 | 1442 |
| Skhul-Qafzeh | 90,000 | 66.6 | 1444 |
| Early Late Pleistocene | 100,000 – 150,000 | 67.7 | 1307 |
| Late Middle Pleistocene | 200,000 – 300,000 | 65.6 | 1148 |
| Middle Middle Pleistocene | 400,000 – 550,000 | 67.9 | 1057 |
| Late Early to Early Middle Pleistocene | 600,000 – 1,150,000 | 58.0 | 835 |
| Early Pleistocene | 1,200,000 – 1,800,000 | 61.8 | 890 |

(Data from Christopher Ruff and his colleagues.)

The invention of cooking increased the nutritional quality and variety of foods available to humans and facilitated digestion. The heat of cooking predigests proteins, thus easing the release of their constituent amino acids and absorption by the gut. Cooking also breaks down toxins in vegetable matter and kills pathogenic microorganisms. In these ways cooking enhanced the quality and quantity of nutrients available for supporting brain tissue. It also would have indirectly increased the energy available for brain metabolism by reducing the energy required for digestion. Thus the invention of cooking would have been an enabling factor in the evolution of large brain size in humans. Archaeological evidence, in the form of hearths and firepits, indicates that the extensive controlled use of fire by humans for warmth and cooking probably arose in the period between 500,000 and 100,000 years ago, when average brain size in humans expanded by 24 percent.

colonized environments that were much more variable in temperature, rainfall, food resources, and risk of predation. The expansion of the brain and its capacity to buffer environmental variation occurred at this time of colonization of new habitats.

*Homo erectus* then entered into a period, lasting more than a million years, during which brain and body size remained stable. During this time, *Australopithecus robustus*, which evidently was specialized for eating tough, fibrous plants, became extinct. Perhaps this diet of difficult-to-digest foods limited brain size in *Australopithecus robustus* and thus contributed to its failure in competition with *Homo erectus*. About half a million years ago, hominid brain and body size again began to increase, and this population gave rise to two descendant

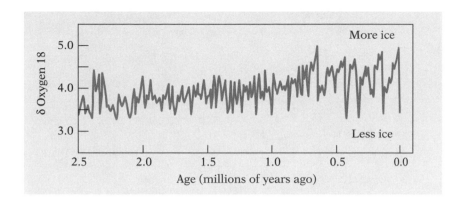

Changes in oxygen isotope concentration as a marker for glaciation over the past 2.5 million years, from the work of Thomas Crowley and Gerald North. The large upward fluctuations in the last 700,000 years corresponds to cycles of massive glaciations during the Pleistocene epoch. The measurements were taken from cores drilled from the ocean floor and reflect the influence of water temperature on the isotope composition of carbonate deposits. In Africa there were related fluctuations from wet to dry conditions with an overall trend toward increasing aridity.

groups, the Neanderthals and the early *Homo sapiens*. This second phase of brain expansion coincided with the very large fluctuations in climate associated with the glacial and interglacial intervals of the Pleistocene epoch during the last 700,000 years. The causes of the Pleistocene climatic changes have not been established, but on average the climate was colder than during any other time in the last 65 million years. For most of the world it was a period of intense cold punctuated by brief warm intervals. Ninety percent of the last 700,000 years was colder than our weather is today. Rick Potts has suggested that abrupt changes probably happened within the span of a human lifetime, and recent high-resolution studies of ice cores from the late Pleistocene by Jeffrey Severinghaus and his colleagues indicate that major changes occurred in less than a decade.

The ancestral stock that gave rise to the hominids and the living great apes probably possessed many features found in orangutans, gorillas, and chimpanzees. These features included the long inter-birth intervals and the very limited paternal care for offspring found in the living great apes. Long interbirth intervals severely limit the fecundity of ape populations and may explain why monkeys and humans have been far more successful both in terms of numbers and geographical distribution, and why the living great apes are restricted to relic populations in Africa and East Asia. Another limiting factor is the older age of great ape mothers, which do not begin to have babies until they are 10 to 15 years old, much later than other nonhuman primate mothers. As shown in the graph on page 176, the age of sexual maturity is a function of relative brain weight. Projecting the regression line for the relationship between sexual maturity and brain-weight residuals, we would expect that humans

| Maturation Stages in Humans | | | |
|---|---|---|---|
| | AGE (YEARS) | | |
| | PREDICTED FROM BRAIN WEIGHT | ACTUAL | RATIO |
| First molar | 19.3 | 6.4 | 0.33 |
| Second molar | 29.2 | 11.1 | 0.38 |
| Wisdom teeth | 37.8 | 20.5 | 0.54 |
| Sexual maturity | 44.5 | 16.6 | 0.37 |
| Maximum life span | 101.5 | 105.0 | 1.03 |

(Based on the graph on page 176.)

would become sexually mature at about age 44, but this is obviously much later than the age of sexual maturity in any human population. In the Aché, a forest-dwelling hunter-gatherer population, the average age of sexual maturity in females is 15; in contemporary urban populations throughout the world it is much earlier. Similarly, the eruption of the third molars, the "wisdom teeth," is another measure of reaching adulthood. The age of wisdom teeth eruption is also strongly related to brain weight. By projecting the regression line for the relationship between the age of wisdom-teeth eruption and brain-weight residuals, we would expect that humans would reach this measure of adulthood at 38 years, when in fact the average age for the eruption of wisdom teeth is only 20.5 in humans. Thus the age of sexual maturity and the age of wisdom-teeth eruption are much earlier in humans than would be expected for a primate with our brain weight. At earlier stages in the developmental cycle, the times of eruption of the first and second molar teeth are also related to relative brain size. All these measures also show a similar early maturation in humans relative to the expected time for a primate of our brain size.

Thus the whole developmental timetable is advanced in humans. If the brain is the pacemaker for development, humans have accelerated maturation relative to what would be expected for a primate of our brain size. This accelerated timetable in humans may be related

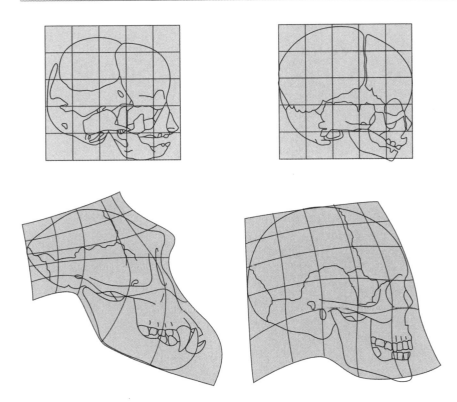

The pedomorphy of the human skull is evident when the growth of the chimpanzee skull (left) and the human skull (right) is plotted on transformed coordinates. The chimpanzee and human skulls are very similar at the fetal stage (top images), but with maturation the chimpanzee skull diverges much more from the fetal pattern than the human.

to the common observation that adult humans resemble immature apes. There is evidence that humans evolved from apes through a process of pedomorphism, the retention of immature qualities of the ancestral group by adults in the descendant group. Pedomorphic features in humans include a large brain relative to body size and a small jaw. Thus human development with respect to nonbrain structures appears to be a truncated version of ape development, with humans becoming reproductive at an immature stage relative to apes. A similar pedomorphic transformation occurred at the origin of mammals.

The accelerated maturation of humans relative to the time expected for our brain size bears on the nature of the relationship between brains and time. I have suggested that brains have evolved to deal with environmental uncertainty and that the longer the life span, the greater the uncertainty. It is also clear that it takes a long time to develop a big brain, and this probably has to do with the slow acquisition of experience necessary for a big brain to function well. Thus a key element in the brain–time relationship may be the time required to reach maturity. However, as is shown in the graph on page 169 and the table on page 196, the maximum life span for humans is almost exactly what would be expected for a primate of our brain size, suggesting that the ultimate variable in the relationship between brains and time is the life span as a whole rather than the time required for development.

The interbirth interval in the human populations tends to be shorter than in the great apes, and thus the reproductive potential of humans is larger than that of the great apes. For example, in Aché women the average interbirth interval is a little less than 3 years, a significantly shorter period than the 4 to 8 years typical of the great apes. In other human populations that do not use birth control the interbirth interval can be even shorter. I believe that this great difference between the demographic structure of great ape and human populations was made possible by the invention of the human family. In great apes, mothers are largely dependent on their own resources to support their slowly developing offspring. In humans, mothers typically have the support of a mate and often an extended family including siblings, parents, and grandparents. Her mate and relatives serve to buffer her from some of the crises that might overwhelm a youthful, inexperienced mother. Thus human mothers are able to reproduce at a much earlier age than would be possible without the family support structure. In turn, earlier maturation enables humans to have a much greater reproductive potential than the great apes. The invention of the extended family enabled humans to evolve much larger brains and avoid the constraints imposed by the extremely slow maturation and low fecundity associated with such large brain size.

In nonhuman primates, paternal care usually occurs in the context of a family unit. Siamangs, owl monkeys, and titi monkeys live in pairs with their dependent offspring. Marmosets and tamarins live in extended families in which there are additional adults that serve as caretakers. The extended family as a social structure developed at some point in hominid evolution. The reduction in the size difference between males and females observed in early humans may be a direct expression of this change in social structure, since a lack of size difference between the sexes is a characteristic feature of family-living primates. For example, siamang males are 8 percent heavier than females, and owl monkey males are only 3 percent larger than their mates. It is significant that the reduction of size difference between the sexes of hominids occurred at about the same time as the brain began to expand. Large-brained infants are very

## Bonding Hormones

Oxytocin and arginine-vasopressin (AVP) are members of an ancient family of hormones that regulate reproduction and other basic physiological functions in both vertebrates and invertebrates. Oxytocin is a chain of nine amino acids made in the hypothalamus and secreted into the bloodstream. It is well known for its roles in female reproductive physiology, promoting the release of milk by the breasts and stimulating uterine contractions during labor. Oxtyocin has a role in establishing the bonds between mother and nursing infant. Oxytocin also reaches peak levels in the bloodstream during orgasm in both sexes and thus may promote bonding between mates. Arginine-vasopressin is also produced by the hypothalamus and consists of a very similar nine-amino-acid chain. Thomas Insel and his colleagues have studied the role of AVP in pair-living versus promiscuous voles. They mapped the distribution of AVP receptors in the brains of prairie voles, which live in pairs, and of montane voles, which are promiscuous. The distribution is quite different in the two species. They have also shown that blocking the action of AVP eliminated the strong preference for a specific mate that is characteristic of the monogamous voles. In pair-living voles, the father assists in the care of infants, and this behavior also depends on AVP. It would be extremely interesting to determine whether pair-bonding and the paternal care of infants in primates similarly depends on AVP.

expensive to nurture and, as Owen Lovejoy has suggested, the recruitment of paternal support may have made it possible to sustain their development in early humans. This change in male behavior may have been brought about by changes in the mechanisms of the hormones oxytocin and arginine-vasopressin, which are associated with paternal care of offspring in mammals. There is also evidence that the age of sexual maturity in humans depends on nutritional status, which is enhanced by the foraging expertise of the extended family. Aché girls who are heavier reach sexual maturity at an earlier age.

There is a direct linkage between body fat and puberty. The hormone leptin is secreted by fat cells and transported in the bloodstream to the hypothalamus, where it binds to leptin receptors. Studies by Karine Clement, A. Strobel, and their colleagues have shown that human females who have mutations that cause a defect in either the production of leptin or in its receptor do not mature sexually because their brains do not receive the signal indicating that the girls have sufficient fat reserves to sustain pregnancy. These results imply that one of the consequences of the improved food supply provided by the extended family would be more rapid growth and earlier sexual maturity.

The human extended family not only shares food, it also shares information, which enhances survival. Vocal communication facilitates the process of sharing information. Pair-living species typically use a rich array of vocalizations to attract mates, enhance pair bonding, and defend their families' home territories against others of their species. Vocalizations are intimately related to this mating system in birds, where 90 percent of the species live in pairs. Vocalizations are also a conspicuous feature of the behavior of pair-living primates like siamangs, owl monkeys, and titi monkeys. Primitive language may have developed in early humans as a means to facilitate family bonding and coordinate food acquisition. There is considerable evidence that gorillas and chimpanzees have some capacity to understand and use language. The evidence for linguistic capacity in apes is especially compelling in the experiments with bonobos done by E. Sue Savage-Rumbaugh and Duane Rumbaugh. Thus the neural apparatus for language-like behavior may have already existed in the common ancestors of apes and humans.

Two areas of the human brain are responsible for speech, both located in the left hemisphere in most people. The first speech area is located in the frontal lobe and was discovered by Paul Broca in 1861. Located just in front of the motor cortex, Broca's area controls the articulation of speech. In monkeys, the zone in front of motor cortex is involved in the visual guidance of motor acts, and here Giacomo Rizzolatti and his colleagues found mirror neurons, which respond when the subject observes a specific motor act performed by another. These neurons may establish a correspondence between seeing an act and performing it. Although Broca's area is classically associated with movements of the vocal apparatus, recent brain-imaging studies have shown that it is also activated by hand movements. Thus Rizzolatti and his colleagues have proposed that Broca's area is involved in matching observed vocal and manual gestures with the production of these same gestures. An interesting possibility is that the mirror cells may be active when an infant learns to mimic speech. The importance of visual input to the interpretation of vocal communication is also shown by the remarkable "McGurk effect," in which watching a person enunciate a speech sound can influence the observer's auditory perception of that sound. For example, the images of a subject pronouncing "ga" presented together with the sound "ba" will override the observer's auditory perception of the acoustic "ba" so that it is perceived as "da."

The human neural control system for the production of speech sounds in Broca's area may be the outgrowth of a more ancient system for the observational guidance of movements and gestures. Ralph Halloway found that the surface impressions left by the brains of *Homo habilis* show an enlargement of the part of the frontal lobe corresponding in location to Broca's area in modern humans. These findings indicate that there exists an area in the frontal lobe of nonhuman primates that may be involved in the mimicry of oral and manual gestures and that this area expanded in the earliest humans.

The second speech area is located in the left temporal lobe and was discovered in autopsies of language-impaired patients by Carl Wernicke in 1874. Wernicke's area is more concerned with the comprehension of language. Joseph Rauschecker and his colleagues have found that part of the comparable region of the temporal lobe in macaque monkeys contains specialized populations of neurons that are involved in social communication. In the lateral auditory areas in the superior temporal gyrus, most neurons are highly sensitive to species-typical vocalizations. Patricia Kuhl and Denise Padden found that macaques have enhanced discrimination capacity at the acoustic boundaries between speech sounds such as the transition between "ba" and "ga" in much the same way humans do. Thus the auditory cortex in nonhuman primates is organized so as to discriminate speech phonemes, even though these animals cannot produce speech sounds. In the bottom of the adjacent superior temporal fissure, there is a region that is responsive to both auditory and visual input. Many of the neurons in this location are exquisitely sensitive to the images of faces, and damage to this region impairs the ability of monkeys to interpret eye contact, an important social signal. In humans the absence of pigment in the sclera serves to highlight the iris and enhance the eye-contact signal. Robert Desimone proposed that these parts of the temporal lobe, which are concerned with multimodal social communication in monkeys, may be related to Wernicke's area in humans.

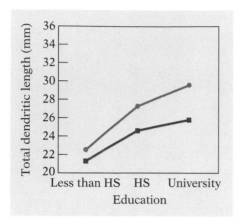

Education and dendritic growth in Wernicke's area in humans, from the work of Bob Jacobs, Matthew Schall, and Arnold Scheibel. The data are taken from measurements of Golgi-stained pyramidal neurons in 20 right-handed subjects. Total dendritic length was measured for 20 randomly selected neurons (red) from each subject. Proximal dendrites are located near the cell body and grow earlier in development; distal dendrites (blue) are located farther from the cell body and are more sensitive to environmental influences. The effects of age were controlled for in these data.

Another line of evidence relevant to the evolution of the neural apparatus for language comes from a recent study by Richard Kay, Matt Cartmill, and Michelle Balow. They reasoned that the size of the nerve controlling the muscles of the tongue is related to the capacity of the tongue to enunciate different speech sounds. The nerve passes through the hypoglossal canal in the skull. They found that the hypoglossal canal is 1.8 times larger in humans than in the apes. The canal is ape-sized in australopithecines, but at least 300,000 years ago it achieved the size found in modern humans, implying that human speech is at least this old.

Taken together, the ape-language studies and the neurobiological studies in monkeys indicate that australopithecines probably had some capacity for comprehending language and the neural circuitry necessary for imitating vocal gestures. The network of family support developed by early humans enabled mothers to reproduce at a younger age than would be expected from their brain size. Without the extended family, big brains would not have evolved in hominids. Language developed as a gestural system that enhanced bonding and the coordination of activities within the extended family. The development of a symbolic system enabled early humans to communicate about objects that were not present, including distant food resources that might be harvested in the future. This symbolic system provided a means for sharing knowledge and coordinating survival strategies in the face of environmental uncertainty. It also facilitated the transmission and accumulation of knowledge from generation to generation. Young apes learn about their environment by observing older animals and by trial and error. Members of human extended families actively *teach* their young about their environment. Teaching accelerates the acquisition of knowledge necessary for adult competence, and it may accelerate the growth of dendrites and the formation of synapses and thus the functional maturation of the brain. In a carefully controlled study, Bob Jacobs, Matthew Schall, and Arnold Scheibel found that the length of dendrites of neurons in Wernicke's area increased according to the amount of education that the subjects had received. The extended family unit, perhaps with a sexual specialization of food-gathering and hunting roles, was able to obtain higher quality and more easily digestible foods, which enabled early humans to devote less energy to digestion and more to the brain. Finally, as the economist Gary Becker pointed out in his *Treatise on the Family*, the extended family "is important in traditional societies in large measure because it protects members against uncertainty." The human evolutionary success story depends on two great buffers against misfortune, large brains and extended families, with each supporting and enhancing the adaptive value of the other. Recent human history is largely related to the development of additional buffers in the form of cultural institutions: governments, churches, and various commercial enterprises such as insurance companies, which protect against uncertainty. Educational institutions enhance brain functions and may directly influence the growth of the dendrites and the formation of connections among neurons in the brain.

Brains originated more than half a billion years ago during a period of extraordinary climatic instability in the early part of the Cambrian period. The expansion of the human brain also appears to have been linked to the challenge of habitat variability. The first major phase of human brain expansion about 2 million years ago coincided with the colonization of drier and more variable habitats and the migration of humans out of Africa into Eurasia. The second major phase of expansion of the human brain occurred during the rapid climatic fluctuations of the Pleistocene glacial and interglacial periods beginning about 700,000 years ago and was associated with the migration of humans into severely cold climates.

The uniquely human feature that differentiates us from other primates is our ability to participate in a large variety of different social networks, each with its own rights and obligations. Nearly all humans participate in a nuclear family, wider webs of kinship, villages, neighborhoods, and a variety of economic, political, and ceremonial relationships. Because we participate in so many different social networks, it is impossible to specify a social group size for humans. Participation in each of these social networks requires a different set of behaviors depending on the context. The adaptive significance of self-awareness in humans is the ability to navigate these different networks with the appropriate behavior for the particular context. Behaviors that are expected in one social context must be withheld in others. Although many parts of the brain probably contribute to this capacity to participate in multiple social networks, the anterior cingulate cortex and connected parts of the frontal lobe appear to be particularly involved in self-control and social awareness, as found by Marcus Raichle, Michael Posner, Richard Lane, and Antonio Damasio.

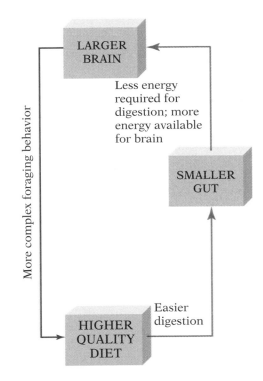

A model for evolutionary changes in diet, brain, and gut in hominids proposed by Leslie Aiello and Peter Wheeler. An important aspect of the more complex foraging behavior was the development of cooperative hunting techniques.

# Dances with Wolves

The wolf pack is held together by strong social bonds maintained through communal vocalizations and elaborate social rituals. The wolf cub is rubbing the older wolf's muzzle, encouraging the older wolf to share food with it. All adult members of the pack assist in the care of the cubs. Intelligence and cooperative hunting techniques enable wolves to be extremely formidable predators.

If a team of ecologists interested in mammals at the top of the food chain were transported back in time 150,000 years, they would find a single highly successful species of wolf, *Canis lupus*, living throughout most of Eurasia and North America. They would also find various forms of humans, each with a much more restricted geographical range. They would find the ancestors of Neanderthals living in the cold climate of Europe and western Asia and perhaps a relic population of *Homo erectus* living in rain forests of southeastern Asia. They would also find in Africa the ancestors of modern humans. Which of these mammals would have been deemed the most successful at that time? My hunch is that our time-traveling ecologists would probably vote for the wolves. Wolves and humans shared much in common. They were highly mobile, cooperative predators that when possible captured ungulates, hoofed mammals such as sheep, but were also opportunistic hunters of smaller prey and would even scavenge the kills of other predators when available. Both wolves and humans lived in highly vocal extended families in which both females and males cared for and provided food for the young. The extended family is a very rare type of social structure in mammals, and I believe it was essential to the success of both wolves and humans.

The geographical distribution of wolves and humans 150,000 years ago. There were wolves living throughout Eurasia and most of North America; the ancestors of Neanderthals lived in Europe and the Near East. The ancestors of modern humans lived in Africa. Other populations of humans may have existed in Asia.

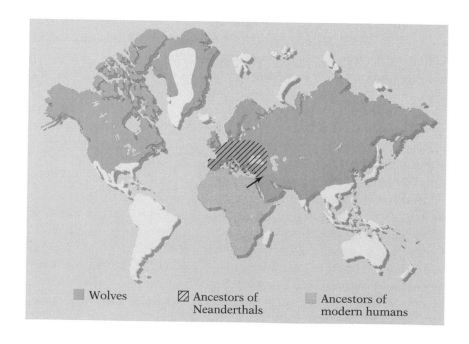

■ Wolves    ▨ Ancestors of Neanderthals    ■ Ancestors of modern humans

Comparative studies of DNA from contemporary humans from throughout the world indicate that the most recent common ancestor of all non-African populations lived in Africa about 140,000 years ago. The migration of the ancestors of humans living outside Africa began at some point after this. Why was this migrating population so successful that it supplanted the other human populations then living in Eurasia? Their success may be because as they entered Asia the migrants encountered wolves and domesticated them. Recently Robert Wayne and his colleagues have done a comprehensive study of mitochondrial DNA sequences from a large number of wolves, dogs, and other canids from throughout the world. They have found that wolves were ancestral to dogs and that the initial domestication of dogs from wolves began as long as 135,000 years ago. This domestication was greatly facilitated because humans and wolves shared similar cooperative hunting behaviors and extended family social structure. Thus wolves and humans were pre-adapted to fit into each other's ecologies and families. Tame wolves and their descendants would have provided an enormous competitive advantage for the human groups that domesticated them. Dogs may have been responsible for the expansion of the human range into Siberia and North America, which wolves already occupied. Dogs would have helped their human companions to obtain food by their strength, stamina, and skills as cooperative hunters. The dog's superior senses of smell and hearing would have complemented the early human's keen vision in detecting both prey and predators. Dogs are more alert at night than are humans and thus better able to detect nocturnal predators. The heritage of close social bonding within the wolf pack and their keen social intelligence allowed dogs to bond with humans and fit in easily as a members of human extended families. Dogs in turn benefited from human assistance in the task of rearing their pups. From the dog's point of view, humans were pack members who brought food to the pups. Human support enabled dogs to have two litters of pups per year instead of the single litter in wolves.

Wolf biologist David Mech and wildlife photographer Jim Brandenburg have made several observations relevant to domestication during their study of the wolves living on Ellesmere Island in the high Canadian Arctic. Continuous summer daylight and the absence of trees and other humans greatly facilitated their observations. They found that the dominant male and female took possession of large prey after it had been killed. The subordinate members of the pack submissively approached the dominant male and licked his face, just as the wolf cubs solicit food from older animals. The

Adult wolves soliciting meat from the dominant male by licking his face after the pack killed a musk ox.

DNA data indicate that the first domestication of wolves occurred as long as 135,000 years ago. Above: A man shares meat with a pup just as wolves share meat with the young in their packs. A dog uses its keen senses to locate prey. Facing: A dog grab-bites prey. Dogs, by virtue of their speed and stamina, are superior to humans in running down and seizing prey. As a joint family of humans and dogs gathers around a camp fire at night, one of the dogs detects and wards off potential intruders. Dogs are also a source of warmth on cold nights. The domestication of wolves conferred an enormous selective advantage on the humans who accomplished it.

Ellesmere wolves had never been persecuted by humans and were intensely curious about their observers. They frequently approached them in a submissive and sometimes even playful manner. In one case a wolf cub untied Mech's bootlaces. I think that the initial domestication came about by raising wolf cubs, which when grown assisted humans in cooperative hunting. Humans, like the dominant pair in the wolf pack, controlled access to the killed prey and distributed portions to submissive grown cubs. The essential act in domestication was maintaining dominance over the wolf cubs once they had matured, the rest was part of their natural behavior. When dogs lick the faces of their human masters it is a direct survival of this ancient food-begging behavior. This theory stands in contrast to the conventional view that the ancestors of dogs became domesticated as they scavenged human refuse. The difficulty with that theory is that many animals scavenge human refuse, but very few have been domesticated. It also does not explain the uniquely close bond between humans and dogs and the widespread use of dogs for many practical applications, such as protection, hunting, and herding.

The remains of wolves have been found together with those of early humans, but it is extremely difficult to determine the initial stages of canine domestication from their bones. Dogs have been found buried in the floors of houses and with humans in graves 12,000 to 14,000 years old at various sites throughout Eurasia.

These burials indicate a close association with humans, but rather than reflecting the initial stage of domestication they may instead reveal the product of a long process leading to this stage. Some of these early dog remains come from settlements with stone buildings that contained mortars and pestles for grinding grain. As Robert Wayne has pointed out, these remains may reflect biological changes in dogs that were induced by a more sedentary human lifestyle and a reduced reliance on hunting for food. Earlier dogs may have very closely resembled wolves. The fact that nearly all modern human groups, including the aboriginal inhabitants of the New World, keep dogs is additional evidence that dogs were domesticated long before the advent of agriculture. The early domestication of dogs is also supported by the great variation among breeds of dogs, which presumably would have taken a long time to evolve.

In the process of domestication, humans have induced a series of pedomorphic changes in dogs that resemble the pedomorphic changes that occurred in human evolution. Some of these changes are anatomical. For example, Pekinese have been bred to resemble infants with their reduced jaws. Selective breeding has also been used to change behavior. The predatory act of wolves involves a series of distinct behaviors—stalking, chasing, grab-biting, and kill-biting—that are acquired in progressive stages as wolves mature. Dog breeders have selected against some of these behaviors. Sheep-herding dogs, for example Border Collies, have been bred to stalk and chase but do not to grab or kill sheep. By contrast, sheep-guarding dogs, for example Komondors, have been bred selectively to eliminate all the predatory behaviors toward sheep. Thus maturation-specific behaviors are under genetic control and can be readily selected for. Dogs probably had an important role in the domestication of sheep and other ungulates. Initially humans and their dogs would have preyed upon sheep, but gradually they shifted their relationship with the sheep by protecting them from other predators and controlling their movements so as to have a readily available supply of meat. About 8000 years ago a mutation occurred in sheep that produced a woolly coat suitable for use in textiles and clothing, which serve as another means of protecting humans from environmental variability.

Dogs have brains that are about three quarters the size of the brains of wolves of comparable body size; the process of domestication leads to a reduction in brain size. Body size is also reduced in most breeds of dogs. In domestication, humans have assumed responsibility for providing food and shelter for dogs, and thus the dog's necessity for maintaining a larger brain is decreased. Similar reductions in brain size and neuron number have been observed in other domestic animals relative to their wild counterparts. One particularly well documented example comes from the observations of Rob Williams and his colleagues, who compared the retinas of European wild cats (*Felis silvestris*) and domestic cats. They found about the same number of photoreceptors in both wild and domestic cats, but a 40 percent reduction in domestic cats in the number of retinal ganglion cells, which relay the retinal input to the brain. They proposed that this major loss of neurons is produced by programmed cell death in development and is linked to domestication.

As Robert Martin has pointed out, human brain size has also decreased over the past 35,000 years. Early modern humans had brains that averaged about 1450 grams, whereas the average for contemporary humans is about 1300 grams. Christopher Ruff and his colleagues have found that this reduction in brain weight was associated with a parallel reduction in body weight. It has been only in the last 100 years or so that body weight increased. The past 35,000 years has been a period of rapid development of every aspect of culture, with an increasing mastery of the physical world, yet ironically it has been associated with a reduction rather than an increase in brain size. The domestication of plants and animals as sources of food and clothing served as major buffers against environmental variability. Perhaps humans, through the invention of agriculture and other cultural means for reducing the hazards of existence, have domesticated themselves.

# READINGS

Adolphs, R., Tranel, D., Damasio, H., and Damasio, A. 1995. Fear and the human amygdala. *Journal of Neuroscience* 15:5879–5891.

Aiello, L., and Wheeler, P. 1995. The expensive-tissue hypothesis. *Current Anthropology* 36:199–221.

Albright, T. 1984. Direction and orientation selectivity of neurons in visual area MT of the macaque. *Journal of Neurophysiology* 52:1106–1130.

Allman, J. 1977. Evolution of the visual system in the early primates. In *Progress in Psychobiology and Physiological Psychology*, ed. J. Sprague and A. Epstein, 1–53. New York: Academic Press.

Allman, J. 1982. Reconstructing the evolution of the brain in primates through the use of comparative neurophysiological and neuroanatomical data. In *Primate Brain Evolution: Methods and Concepts*, ed. E. Armstrong and D. Falk, 13–28. New York: Plenum Press.

Allman, J., and Kaas, J. 1971. A representation of the visual field in the caudal third of the middle temporal gyrus of the owl monkey (*Aotus trivirgatus*). *Brain Research* 31: 84–105.

Allman, J., Miezin, F., and McGuinness, E. 1985. Stimulus-specific responses from beyond the classical receptive field: neurophysiological mechanisms for local-global comparisons in visual neurons. *Annual Review of Neuroscience* 8:407–430.

Allman, J., Rosin, A., Kumar, R., and Hasenstaub, A. 1998. Parenting and survival in anthropoid primates: caretakers live longer. *Proceedings of the National Academy of Sciences* 95:6866–6869.

Appel, T. 1987. *The Cuvier-Geoffroy Debate*. New York: Oxford University Press.

Bell, A., and Sejnowski, T. 1997. The "independent components" of natural scenes are edge filters. *Vision Research* 37:3327–3338.

Berg, H. 1988. A physicist looks at bacterial chemotaxis. *Cold Spring Harbor Symposium in Quantitative Biology* 53:1–9.

Brasier, M., Rozanov, A., Zhuravlev, A., Corfield, R., and Derry, L. 1994. A carbon isotope reference scale for the lower Cambrian succession in Siberia. *Geology Magazine* 131:767–783.

Britten, K., Newsome, W., Shadlen, M., Celebrini, S., and Movshon, J. 1996. A relationship between behavioral choice and the visual responses of neurons in macaque MT. *Visual Neuroscience* 13:87–100.

Brownell, W., Bader, C., Bertran, D., and deRibaupierrre, Y. 1985. Evoked mechanical responses of isolated cochlear hair cells. *Science* 227:194–196.

Bruce, L., and Neary, T. 1995. The limbic system of tetrapods. *Brain, Behavior and Evolution* 46: 224–234.

Brunelli, S., Faiella, A., Capra, V., Nigro, V., Simeone, A., Cama, A., and Boncinelli, E. 1996. Germline mutations in the homeobox gene *Emx-2* in patients with severe schizencephaly. *Nature Genetics* 12:94–96.

Bullock, T., and Horridge, A. 1965. *Structure and Function in the Nervous Systems of Invertebrates*. New York: W. H. Freeman.

Carroll, R. 1988. *Vertebrate Paleontology and Evolution*. New York: W. H. Freeman.

Cartmill, M. 1972. Arboreal adaptations and the origin of the order primates. In *The Functional and Evolutionary Biology of Primates*, ed. R. Tuttle, 97–212. Chicago: Aldine-Atherton Press.

Catania, K., and Kaas, J. 1997. Somatosensory fovea in the star-nosed mole: behavioral use of the star in relation to innervation patterns and cortical representation. *Journal of Comparative Neurology* 387:215–223.

Cherniak, C. 1995. Neural component placement. *Trends in Neuroscience* 18:522–527.

Chivers, D. 1974. The siamang in Malaysia. *Contributions to Primatology* 4:1–335.

Clément, K., Vaisse, C., Lahlou, N., Cabrol, S., Pelloux, V., Cassuto, D., Gourmelen, M., Dina, C., Chambaz, J.,

Lacourte, J.-M., Basdavant, A., Bougnères, P., Lebouc, Y., Frogeul, P., and Guy-Grand, B. 1998. A mutation in the human leptin receptor gene causes obesity and pituitary dysfunction. *Nature* 392: 398–401.

Cohen-Tannoudji, M., Babinet, C., and Wassef, M. 1994. Early determination of a mouse somatosensory cortex marker. *Nature* 368:460–463.

Cooper, J., Bloom, F., and Roth, R. 1991. *The Biochemical Basis of Neuropharmacology.* New York: Oxford University Press.

Damasio, A. 1994. *Descartes' Error: Emotion, Reason, and the Human Brain.* New York: Grosset-Putnam.

Damasio, A., Yamada, T., Damasio, H., Corbett, J., and McKee, J. 1980. Central achromatopsia: behavioral, anatomic, and physiologic aspects. *Neurology* 30: 1064–1071.

DeRobertis, E., and Sasai, Y. 1996. A common plan for dorsoventral patterning in bilateria. *Nature* 380:37–40.

Desimone, R., Albright, T., Gross, C., and Bruce, C. 1984. Stimulus-selective properties of inferior temporal neurons in the macaque. *Journal of Neuroscience* 4:2051–2062.

Desmond, A., and Moore, J. 1991. *Darwin: The Life of a Tormented Evolutionist.* New York: Warner Books.

DiDonato, C., Chen, X., Noya, D., Korenberg, J., Nadeau, J., and Simard, L. 1997. Cloning, characterization and copy number of the murine *survival motor neuron* gene: homolog of the *spinal muscular atrophy-determining* gene. *Genome Research* 7:339–352.

Dobbins, A., Jeo, R., Fiser, J., and Allman, J. 1998. Distance modulation of neural activity in the visual cortex. *Science* 281:552–555.

Duboule, D., ed. 1994. *Guidebook to the Homeobox Genes.* Oxford: Oxford University Press.

Duboule, D. 1994. Temporal colinearity and the phylotypic progression: a basis for the stability of the vertebrate Bauplan and the evolution of morphologies through heterochrony. *Development Supplement* 134–142.

Dräger, U., and Hubel, D. 1975. Responses to visual stimulation and relationship between visual, auditory, and somatosensory inputs in mouse superior colliculus. *Journal of Neurophysiology* 38:690–713.

Dubner, R., and Zeki, S. 1971. Response properties and receptive fields of cells in an anatomically defined region of the superior temporal sulcus in the monkey. *Brain Research* 35:528–532.

Dunbar, R. 1998. The social-brain hypothesis. *Evolutionary Anthropology* 6:178–190.

Dyke, B., Gage, T., Alford, P., Swenson, B., and Williams-Blangero, S. 1995. A model life table for chimpanzees. *American Journal of Primatology* 37:25–37.

Epstein, R., and Kanwisher, N. 1998. A cortical representation of the local visual environment. *Nature* 392: 598–601.

Fay, R. 1988. *Hearing in Vertebrates.* Winnetka, Ill.: Hill-Fay.

Fernald, R. 1997. The evolution of eyes. *Brain, Behavior and Evolution* 50:53–259.

Fraser, S., Keynes, R., and Lumsden, A. 1990. Segmentation in the chick embryo is defined by cell lineage restrictions. *Nature* 344:431–435.

Gagneux, P., Boesch, C., and Woodruff, D. 1998. Female reproductive strategies, paternity, and community structure in wild West African chimpanzees. *Animal Behavior.* In press.

Gans, C., and Northcutt, G. 1983. Neural crest and the origin of the vertebrates: a new head. *Science* 220: 268–274.

Garber, P. 1997. One for all and breeding for one: cooperation and competition as a tamarin reproductive strategy. *Evolutionary Anthropology* 5:187–199.

Garcia-Fernandez, J., and Holland, P. 1994. Archetypal organization of the amphioxus *Hox* gene cluster. *Nature* 370:563–566.

Gerhart, J., and Kirschner, M. 1997. *Cells, Embryos, and Evolution.* Malden, Mass.: Blackwell Science.

Glickstein, M., and Whitteridge, D. 1987. Tatsuji Inouye and the mapping of the visual fields on the cerebral cortex. *Trends in Neuroscience* 10:350–353.

Goethe, J. W. von. 1989. *Goethe's Botanical Writings.* Trans. Bertha Mueller. Woodbridge, Conn.: Ox Bow Press.

Golomb, B. 1998. Cholesterol and violence. *Annals of Internal Medicine* 128:478–487.

Goodall, J. 1986. *The Chimpanzees of Gombe.* Cambridge: Harvard University Press.

Gould, S. 1977. *Ontogeny and Phylogenyy.* Cambridge: Harvard University Press.

Gross, C. 1998. *Brain, Vision, Memory.* Cambridge: MIT Press.

Grotwiel, M., Beck, C., Wu, K., Zhu, X., and Davis, R. 1998. Integrin-mediated short-term memory in *Drosophila*. *Nature* 391:455–460.

Hadjikhani, N., Liu, A., Dale, A., Cavanagh, P., and Tootell, R. 1998. Retinotopy and color sensitivity in human visual cortical area V8. *Nature Neuroscience* 1:235–241.

Hakeem, A., Sandoval, G., Jones, M., and Allman, J. 1996. Brain and life span in primates. In *Handbook of the Psychology of Aging*, ed. J. Birren and W. Schaie. San Diego: Academic Press.

Hatini, V., Tao, W., and Lai, E. 1994. Expression of winged helix genes, *BF-1* and *BF- 2*, define adjacent domains within the developing forebrain and retina. *Journal of Neurobiology* 25:1293–1309.

Hauser, M. 1996. *The Evolution of Communication*. Cambridge: MIT Press.

Herrada, G., and Dulac, C. 1997. A novel family of putative pheromone receptors in mammals with a topographically organized and sexually dimorphic distribution. *Cell* 90:763–773.

Heywood, C., and Cowey, A. 1992. The role of the "face-cell" area in the discrimination and recognition of faces by monkeys. *Philosophical Transactions of the Royal Society of London*, B, 335:31–38.

Hill, K., and Hurtado, M. 1996. *Aché Life History: The Ecology and Demography of a Foraging People*. New York: Aldine de Gruyter.

Holland, N., and Holland, L. 1993. Serotonin-containing cells in the nervous system and other tissues during ontogeny of a lancelet, *Branchiostoma floridae*. *Acta Zoologica* (Stockholm) 74:195–204.

Hubel, D. 1988. *Eye, Brain, and Vision*. New York: Scientific American Library.

Hughlings Jackson, J. 1931. *The Selected Writings of John Hughlings Jackson*. Ed. J. Taylor. London: Hodder and Stoughton.

Ingram, V. 1963. *The Hemoglobins in Genetics and Evolution*. New York: Columbia University Press.

Jacobs, B. 1994. Serotonin, motor activity, and depression-related disorders. *American Scientist* 82:456–463.

Jacobs, G., Neitz, M., and Neitz, J. 1995. Why bushbabies lack color vision. *American Journal of Primatology* 36:130.

Janvier, P. 1996. *Early Vertebrates*. Oxford: Clarendon Press.

Julesz, B. 1971. *Foundations of Cyclopean Perception*. Chicago: University of Chicago Press.

Kaas, J., Nelson, R., Sur, M., Lin, C., and Merzenich, M. 1979. Multiple representations of the body within the primary somatosensory cortex of primates. *Science* 204:521–523.

Kandel, E., Schwartz, J., and Jessell, T. 1991. *Principles of Neural Science*. 3d ed. New York: Elsevier.

Kaplan, J., Potvin Klein, K., and Manuck, S. 1997. Cholesterol meets Darwin: public health and the evolutionary implications of the cholesterol-serotonin hypothesis. *Evolutionary Anthropology* 6:28–37.

Karten, H., Hodos, W., Nauta, W., and Revzin, A. 1973. Neural connections of the "visual wulst" of the avian telencephalon. *Journal of Comparative Neurology* 150: 253–278.

Kay, R., Cartmill, M., and Balow, M. 1998. The hypoglossal canal and the origin of human vocal behavior. *Proceedings of the National Academy of Sciences* 95: 5417–5419.

Kemp, T. 1982. *Mammal-like Reptiles and the Origin of Mammals*. London: Academic Press.

Kirschvink, J., Ripperdan, R., and Evans, D. 1997. Evidence for a large-scale reorganization of early Cambrian continental masses by inertial interchange true polar wanderer. *Science* 277:541–545.

Knudsen, E. 1982. Auditory and visual maps of space in the optic tectum of the owl. *Journal of Neuroscience* 2:1177–1194.

Kornack, D., and Rakic, P. 1998. Changes in cell-cycle kinetics during the development and evolution of primate neocortex. *Proceedings of the National Academy of Sciences* 95:1242–1246.

Krubitzer, L. 1995. The organization of neocortex in mammals: are species differences really so different? *Trends in Neuroscience* 18:408–417.

Kuhl, P., and Padden, D. 1983. Enhanced discriminability at the phonetic boundaries for the place feature in macaques. *Journal of the Acoustical Society of America* 73:1003–1010.

Kuida, K., Haydar, T., Kuan, C.-Y., Gu, Y., Taya, C., Karasuyama, H., Su, M., Rakic, P., and Flavell, R. 1998. Reduced apoptosis and cytochrome-C–mediated caspase activation in mice lacking *Caspase-9*. *Cell* 94: 325–337.

Lacalli, T. 1996. Frontal eye circuitry, rostral sensory pathways, and brain organization in amphioxus larvae. *Philosophical Transactions of the Royal Society of London*, B, 351:243–263.

Lacalli, T., Holland, N., and West, J. 1994. Landmarks in the anterior central nervous system of amphioxus larvae. *Philosophical Transactions of the Royal Society of London*, B, 344:165–185.

Leibowitz, H. 1971. Sensory, learned, and cognitive mechanisms of size perception. *Annals of the New York Academy of Sciences* 188:47–62.

Lesch, K.-P. 1998. Serotonin transporter and psychiatric disorders: listening to the gene. *The Neuroscientist* 4:25–34.

Lewis, E. 1992. Clusters of master control genes regulate the development of higher organisms. *Journal of the American Medical Association* 267:1524–1531.

Livingstone, M., and Hubel, D. 1984. Anatomy and physiology of a color system in the primate visual cortex. *Journal of Neuroscience* 4:309–356.

Lovejoy, O. 1981. The origin of man. *Science* 211:341–350.

Lucas, A., Morley, R., Cole, T., Lister, G., and Leeson-Payne, C. 1992. Breast milk and subsequent intelligence quotient in children born preterm. *Lancet* 339:261–264.

Margulis, L. 1992. *Symbiosis in Cell Evolution*. 2d ed. New York: W. H. Freeman.

Martin, R. 1990. *Primate Origins and Evolution: A Phylogenetic Reconstruction*. Princeton: Princeton University Press.

Matsunami, H., and Buck, L. 1997. A multigene family encoding a diverse array of putative pheromone receptors in mammals. *Cell* 90:775–784.

McCaffery, P., Mey, J., and Dräger, U. 1996. Light-mediated retinoic acid production. *Proceedings of the National Academy of Science* 93:12570–12574.

McGuinness, E., Sivertsen, D., and Allman, J. 1980. Organization of the face representation in macaque motor cortex. *Journal of Comparative Neurology* 193:591–608.

McHenry, H. 1994. Behavioral ecological implications of early hominid body size. *Journal of Human Evolution* 27:77–87.

McMenamin, M., and McMenamin, D. 1990. *The Emergence of Animals*. New York: Columbia University Press.

Meyer, A. 1998. *Hox* gene variation and evolution. *Nature* 391:225–228.

Mollon, J. 1995. Seeing colour. In *Colour: Art and Science*, ed. T. Lamb and J. Bourriau, 127–150. Cambridge: Cambridge University Press.

Monaghan, P., Grau, E., Bock, D., and Schutz, G. 1995. The mouse homologue of the orphan receptor *tailless* is expressed in the developing forebrain. *Development* 121:839–853.

Nathans, J. 1989. Genes for color vision. *Scientific American* 260:42–49.

Ngai, J., Dowling, M., Buck, L., Axel, R., and Chess, A. 1993. The family of genes encoding odorant receptors in the channel catfish. *Cell* 72:657–666.

Nieuwenhuys, R., Ten Donkelaar, H., and Nicholson, C. 1998. *The Central Nervous System of Vertebrates*. New York: Springer-Verlag.

Nilsson, G. 1996. Brain and body oxygen requirements of *Gnathonemus petersii*, a fish with an exceptionally large brain. *Journal of Experimental Biology* 199:603–607.

Nishida, T. 1990. *The Chimpanzees of the Mahale Mountains*. Tokyo: University of Tokyo Press.

Novacek, M. 1992. Mammalian phylogeny: shaking the tree. *Nature* 356:121–125.

Nudo, R., and Masterton, B. 1990. Descending pathways to the spinal cord. III: Sites of origin of the corticospinal tract. *Journal of Comparative Neurology* 296:559–583.

Nudo, R., Plautz, E., and Milliken, G. 1997. Adaptive plasticity in primate motor cortex as a consequence of behavioral experience and neuronal injury. *Seminars in Neuroscience* 9:13–23.

Okun, E. 1970. The effect of environmental temperature on the production of ultrasounds by isolated non-handled albino mouse pups. *Journal of Zoology* 162:71–83.

O'Leary, D., Schlaggar, B., and Stanfield, B. 1992. The specification of sensory cortex: lessons from cortical transplantation. *Experimental Neurology* 115:121–126.

Olesiuk, P., Bigg, M., and Ellis, G. 1988. Life history and population dynamics of resident killer whales (*Orcinus orca*) in the coastal waters of British Columbia and Washington state. *Report of the International Whaling Commission* 12:209–243.

Osorio, D., and Vorobyev, M. 1996. Colour vision as an adaptation to frugivory in primates. *Proceedings of the Royal Society of London*, B, 263:593–599.

Pecins-Thompson, M., Brown, N., and Bethea, C. 1998. Regulation of serotonin re-uptake transporter mRNA expression by ovarian steroids in rhesus monkeys. *Molecular Brain Research* 53:120–129.

Pelligrino, G., Fadiga, L., Fogassi, L., Gallese, V., and Rizzolatti, G. 1992. Understanding motor events: a neurophysiological study. *Experimental Brain Research* 91:176–180.

Peroutka, S., and Howell, T. 1994. Molecular evolution of G protein–coupled receptors. *Neuropharmacology* 3:319–324.

Perrett, D., Smith, D., Potter, P., Mistlin, A., Head, A., Milner, A., and Jeeves, M. 1985. Visual cells in the temporal cortex sensitive to face view and gaze direction. *Proceedings of the Royal Society of London,* B, 223:293–317.

Peters, A., Palay, S., and Webster, H. 1991. *The Fine Structure of the Nervous System.* New York: Oxford University Press.

Petersen, S., Baker, J., and Allman, J. 1980. Dimensional selectivity of neurons in the dorsolateral visual area of the owl monkey. *Brain Research* 197:507–511.

Petersen, S., Baker, J., and Allman, J. 1985. Direction-specific adaptation in area MT of the owl monkey. *Brain Research* 346:146–150.

Petersen, S., Miezin, F., and Allman, J. 1988. Transient and sustained responses in four extrastriate visual areas of the owl monkey. *Experimental Brain Research* 70:55–60.

Pettigrew, J. 1979. Binocular visual processing in the owl's telencephalon. *Proceedings of the Royal Society of London,* B, 204:436–454.

Pettigrew, J. 1986. Flying primates? Megabats have the advanced pathway from eye to midbrain. *Science* 231:1304–1306.

Polyak, S. 1957. *The Vertebrate Visual System.* Chicago: University of Chicago Press.

Potts, R. 1996. *Humanity's Descent: The Consequences of Ecological Instability.* New York: Avon Books.

Qiu, M., Anderson, S., Chen, S., Meneses, J., Hevner, R., Kuwana, E., Pedersen, R., and Rubenstein, J. 1996. Mutation of the *Emx-1* homeobox gene disrupts the corpus callosum. *Developmental Biology* 178:174–178.

Quartz, S., and Sejnowski, T. 1997. The neural basis of cognitive development. *Behavioral and Brain Sciences* 20:537–596.

Quiring, R., Walldorf, U., Kloter, U., and Gehring, W. 1994. Homology of the *eyeless* gene of *Drosophila* to the *small eye* gene in mice and *aniridia* in humans. *Science* 265:785–789.

Raleigh, M., McGuire, M., Brammer, G., Pollack, D., and Yuwiler, A. 1991. Serotonergic mechanisms promote dominance acquisition in adult male vervet monkeys. *Brain Research* 559:181–190.

Raleigh, M., McGuire, M., Melega, W., Cherry, S., Huang, S.-C., and Phelps, M. 1995. Neural mechanisms supporting successful social decisions in simians. In *Neurobiology of Decision-Making,* ed. A. Damasio, H. Damasio, and Y. Christen, 63–82. New York: Springer-Verlag.

Ralph, M., Foster, R., Davis, F., and Menaker, M. 1990. Transplanted suprachiasmatic nucleus determines circadian rhythm. *Science* 247:975–978.

Rauschecker, J., Tian, B., and Hauser, M. 1995. Processing of complex sounds in the macaque nonprimary auditory cortex. *Science* 268:111–114.

Rosa, M., and Schmid, L. 1994. Topography and extent of visual-field representation in the superior colliculus of the megachiropteran *Pteropus. Visual Neuroscience* 11:1037–1057.

Ruff, C., Trinkaus, E., and Holliday, T. 1997. Body mass and encephalization in Pleistocene *Homo. Nature* 387:173–176.

Rupke, N. 1994. *Richard Owen: Victorian Naturalist.* New Haven: Yale University Press.

Salzman, D., Murasugi, C., Britten, K., and Newsome, W. 1992. Microstimulation in visual area MT: effects on direction discrimination performance. *Journal of Neuroscience* 12:2331–2355.

Sapolsky, R. 1992. *Stress, the Aging Brain, and the Mechanisms of Neuron Death.* Cambridge: MIT Press.

Schiller, P. 1993. The effects of V4 and middle temporal (MT) area lesions on visual performance in the rhesus monkey. *Visual Neuroscience* 10:717–746.

Schiller, P., and Stryker, M. 1972. Single-unit recording and stimulation in superior colliculus of the alert rhesus monkey. *Journal of Neurophysiology* 35: 915–924.

Scott, S., Young, A., Calder, A., Hellawell, D., and Aggleton, J. 1997. Impaired auditory recognition of fear and anger following bilerateral amygdala lesions. *Nature* 385:254–257.

Sereno, M., Dale, A., Reppas, J., Kwong, K., Belliveau, J., Brady, T., Rosen, B., and Tootell, R. 1995. Borders of multiple visual areas in humans revealed by functional magnetic resonance imaging. *Science* 268: 889–893.

Shadlen, M., and Newsome, W. 1996. Motion perception: seeing and deciding. *Proceedings of the National Academy of Sciences* 93:628–633.

Shawlot, W., and Behringer, R. 1995. Requirement for *Lim-1* in head-organizer function. *Nature* 374: 425–430.

Smith, H., Crumett, T., and Brandt, K. 1994. Ages of eruption of primate teeth. *Yearbook of Physical Anthropology* 37:177–231.

Smith Fernandez, A., Pieau, C., Repérant, J., Boncinelli, E., and Wassel, M. 1998. Expression of the *Emx-1* and *Dlx-1* homeobox genes define three molecularly distinct domains in the telencephalon of mouse, chick, turtle, and frog embryos: implications for the evolution of telencephalic subdivisions in amniotes. *Development* 125:2099–2111.

Spoont, M. 1992. Modulatory role of serotonin in neural information processing: implications for human psychopathology. *Psychological Bulletin* 112:330–350.

Spudich, J. 1993. Color-sensing in the *Archaea:* a eukaryotic-like receptor coupled to a prokaryotic transducer. *Journal of Bacteriology* 175:7755–7761.

Stoneking, M., and Soodyall, H. 1996. Human evolution and the mitochondrial genome. *Current Opinion in Genetics and Development* 6:731–736.

Striedter, G. 1997. The telencephalon of tetrapods in evolution. *Brain, Behavior and Evolution* 49:179–213.

Swaab, D., Hofman, M., Mirmiran, M., Ravid, R., and Van Leewen, F., eds. 1992. *The Human Hypothalamus in Health and Disease.* New York: Elsevier.

Synder, L., Batista, A., and Andersen, R. 1997. Coding of intention in the posterior parietal cortex. *Nature* 386:167–170.

Thiele, A., Vogelsang, M., and Hoffmann, K.-P. 1991. Pattern of retinotectal projection in the megachiropteran bat *Rousettus aegyptiacus. Journal of Comparative Neurology* 314:671–683.

Thomas, B., Tucker, A., Qui, M., Ferguson, C., Hardcastle, Z., Rubenstein, J., and Sharpe, P. 1998. Role of *Dlx-1* and *Dlx-2* genes in patterning of the murine dentition. *Development* 124:4811–4818.

Tishkoff, S., Deitsch, E., Speed, W., Pakstis, A., Kidd, J., Cheung, K., Bonne-Tamir, B., Santachaira-Benerecetti, A., Moral, P., and Krings, M. 1996. Global patterns of linkable disequilibrium at the CD4 locus and modern human origins. *Science* 271:1380–1387.

Tootell, R., Reppas, J., Dale, A., Malach, R., Born, R., Brady, T., Rosen, B. 1995. Visual motion aftereffect in human cortical area MT/V5 revealed by functional magnetic resonance imaging. *Nature* 375:139–14.

Vanegas, H. 1984. *Comparative Neurology of the Optic Tectum.* New York: Plenum Press.

Van Essen, D. 1997. A tension-based theory of morphogenesis and compact wiring in the central nervous system. *Nature* 385:313–318.

Vila, C., Savolainen, P., Maldonado, J., Amorim, I., Rice, J., Honeycutt, R., Crandall, K., Lundberg, J., and Wayne, R. 1997. Multiple and ancient origins of the domestic dog. *Science* 276:1687–1689.

Welker, W., and Seidenstein, S. 1959. Somatic sensory representation in the cerebral cortex of the raccoon (*Procyon lotor*). *Journal of Comparative Neurology* 111:469–501.

Wells, M. 1962. *Brain and Behavior in Cephalopods.* Stanford: Stanford University Press.

West, G., Brown, J., and Enquist, B. 1997. A general model for the origin of allometric scaling laws in biology. *Science* 276:122–126.

Wicht, H., and Northcutt, G. 1992. The forebrain of the Pacific Hagfish: a cladistic reconstruction of the ancestral craniate forebrain. *Brain, Behavior and Evolution* 40:25–64.

Witman, G. 1993. *Chlamydomonas phototaxis. Trends in Cell Biology* 3:403–408.

Xuan, S., Baptista, C., Baias, G., Tao, W., Soares, V., and Lai, E. 1995. Winged helix transcription factor *BF-1* is essential for the development of the cerebral hemispheres. *Neuron* 14:1141–1152.

Yoshihara, Y., Oka, S., Nemoto, Y., Watanabe, Y., Nagata, S., Kagamiyama, H., and Mori, K. 1994. An ICAM-related neuronal glycoprotein, telencephalin, with brain segment-specific expression. *Neuron* 12: 541–553.

Young, J. 1971. *The Anatomy of the Nervous System of Octopus vulgaris.* Oxford: Clarendon Press.

Young, L., Wang, Z., and Insel, T. 1998. Neuroendocrine bases of monogamy. *Trends in Neuroscience* 21:71–75.

Yuasa, J., Hirano, S., Yamagata, M., and Noda, M. 1996. Visual projection map specified by topographic expression of transcription factors in the retina. *Nature* 382:632–635.

# SOURCES OF ILLUSTRATIONS

Drawings by Joyce A. Powzyk; diagrams and graphs by Fine Line Illustrations.

## CHAPTER 1

Opening image: © Biophoto Associates/Science Source/ Photo Researchers, Inc.

5: Adapted from N. Wade, 10 March 1998, The struggle to decipher human genes, *The New York Times*, F1.

6: Scott Camazine/S. Best, Photo Researchers, Inc.

8: Table derived from H. C. Berg, 1988, A physicist looks at bacterial chemotaxis, *Cold Spring Harbor Symposium in Quantitative Biology* 53:1–9.

10: Tim Flach/Tony Stone Images.

## CHAPTER 2

Opening image: Kenneth Garrett/National Geographic Image Collection.

17: From R. Bauchot, J. Randall, J.-M. Ridet, and M.-L. Bauchot, 1989, Encephalization in tropical teleost fishes and comparison with their mode of life, *Journal für Hirnforschung* 30:645–669.

18: Adapted from unpublished dissections of fish by Brian Rasnow.

20, 23, 24: B. Jacobs, 1914, Serotonin, motor activity, and depression-related disorders, *American Scientist* 82: 456–463.

22: From S. Peroutka and T. Howell, 1994, Molecular evolution of G protein–coupled receptors, *Neuropharmacology* 3:319–324.

29, 39: Courtesy of M. I. Sereno and A. M. Dale.

30: F. Gall and J. Spurzheim, 1810, *Anatomie et Physiologie du Système Nerveux*.

31: Wellcome Institute Library, London.

32: Royal College of Physicians, London.

33: A. S. F. Grünbaum and C. S. Sherrington, 1902, Observations on the physiology of the cerebral cortex of some of the higher apes, *Proceedings of the Royal Society* 69:206–209.

34: Courtesy of Wally Welker.

35: Adapted from photographs provided by Wally Welker and from W. Welker and S. Seidenstein, 1959, Somatic sensory representation in the cerebral cortex of the raccoon (*Procyon lotor*), *Journal of Comparative Neurology* 111:469–501.

36: Kenneth C. Catania and Jon H. Kaas, 1995, Organization of the star-nosed mole, *Journal of Comparative Neurology* 351:549–567. Copyright 1955 Wiley-Liss, Inc.

38: Adapted from J. Allman and J. Kaas, 1976, Representation of the visual field on the medial wall of the occipital-parietal cortex in the owl monkey (*Aotus trivirgatus*), *Science* 191:572–575.

## CHAPTER 3

Opening image: Lennart Nilsson/Albert Bonniers Förlag AB; Lennart Nilsson, 1990, *A Child Is Born* (New York: Dell Publishing).

45: Art Resource, New York.

46: *Cartoon Portraits and Biographical Sketches of Men of the Day: The Drawings of Frederick Waddy*, 2d ed. (London: Tinsley Brothers, 1874).

47 top: Wellcome Institute Library, London.

49: Courtesy of Edward B. Lewis.

50: top, Courtesy of the Archives, California Institute of Technology; bottom, courtesy of Edward B. Lewis.

51: Courtesy of the Archives, California Institute of Technology.

52: E. Lewis, 1992, Clusters of master control genes regulate the development of higher organisms, *Journal of the American Medical Association* 267:1524–1531.

53–54: From A. Feess-Higgins and J.-C. Larroche, 1987, *Le développement du cerveau fœtal humain/Development of the Human Fœtal Brain* (Paris: Editions INSERM/ Masson), 15.

55: R. Keynes and A. Lumsden, 1990, Segmentation and the origin of regional diversity in the vertebrate central nervous system, *Neuron* 4:1–9.

58: S. Xuan, C. Baptista, G. Baias, W. Tao, V. Soares, and E. Lai, 1995, Winged helix transcription factor *BF-1* is essential for the development of the cerebral hemispheres, *Neuron* 14:1141–1152.

61: K. Kuida, T. Haydar, C.-Y. Kuan, Y. Gu, C. Taya, H. Karasuyama, M. Su, P. Rakic, and R. Flavell, 1998, Reduced apoptosis and cytochrome-C–mediated caspase activation in mice lacking *Caspase-9*, *Cell* 94:325–337.

CHAPTER 4

Opening image: Photomicrograph courtesy of Susan Amara, Vollum Institute, Portland, Oregon; produced by Eva Shannon. Cover image, *Journal of Neuroscience*, 15 January 1998.

64, 65: Adapted from S. Gould, 1989, *Wonderful Life* (New York: W. W. Norton).

66: Adapted from J. Kirschvink, R. Ripperdan, and D. Evans, 1997, Evidence for a large-scale reorganization of early Cambrian continental masses by inertial interchange true polar wanderer, *Science* 277:541–545, and M. Magaritz, J. Kirschvink, A. Latham, Zhuravlev, and A. Rozanov, 1991, Precambrian/Cambrian boundary problem: carbon isotope correlations for Vendian and Tommotian time between Siberia and Morocco, *Geology* 19:847–850.

67: Adapted from R. Fernald, 1997, The evolution of eyes, *Brain, Behavior and Evolution* 50:253–259.

69: Adapted from T. Lacalli, N. Holland, and J. West, 1994, Landmarks in the anterior central nervous system of amphioxus larvae, *Philosophical Transactions of the Royal Society of London*, B, 344:165–185.

70: Adapted from T. Lacalli, 1996, Frontal eye circuitry, rostral sensory pathways, and brain organization in amphioxus larvae, *Philosophical Transactions of the Royal Society of London*, B, 351:243–263.

71: Adapted from P. Janvier, 1996, *Early Vertebrates* (Oxford: Clarendon Press), 2, 86.

72: Adapted from A. Meyer, 1998, *Hox* gene variation and evolution, *Nature* 391:225–228.

74: S. Gilbert, 1997, *Developmental Biology* (Sunderland, Mass.: Sinauer), 255.

78: Adapted from R. Nieuwenhuys and C. Nicholson, 1969, Aspects of the histology of the cerebelllum of mormyrid fish, *Neurobiology of Cerebellar Evolution*, ed. R. Llinas (Chicago: American Medical Association), 135–169.

79: Adapted from S. Gilbert, 1997, *Developmental Biology* (Sunderland, Mass.: Sinauer), 279.

80: left, image by Richard Owen, reproduced by kind permission of the President and Council of the Royal College of Surgeons of England; right, © W. E. Townsend, Jr., in J.-Y. Cousteau, 1973, *Octopus and Squid*, trans. P. Diolé (Garden City, N.Y.: Doubleday), 238.

81, 82: Adapted from J. Young, 1971, *The Anatomy of the Nervous System of Octopus vulgaris* (Oxford: Clarendon Press), 6, 424, 448.

CHAPTER 5

Opening image: © Mitsuaki Iwago/Minden Pictures.

87: Adapted from R. Carroll, 1988, *Vertebrate Paleontology and Evolution* (New York: W. H. Freeman).

88, 90–91, 98: Adapted from R. Savage and M. Long, 1986, *Mammal Evolution* (New York: Facts on File).

95: Adapted from T. Kemp, 1982, *Mammal-like Reptiles and the Origin of Mammals* (London: Academic Press), 249.

96: From A. Brink, 1958, Note on a new skeleton of *Thrinaxadon liorhinus*, *Paleontologia Africana* 6:15–22.

99: top, © Ken G. Preston-Mafham/Animals Animals; bottom, adapted from S. Gould, ed., 1993, *The Book of Life* (New York: W. W. Norton), 96.

100: top, from R. Fay, 1988, *Hearing in Vertebrates: A Psychophysics Databook* (Winnetka, Ill.: Hill-Fay; bottom, A. Ladhams and J. Pickles, 1996, Morphology of the monotreme organ of Corti, *Journal of Comparative Neurology* 366:335–347.

101: Adapted from E. Kandel, J. Schwartz, and T. Jessell, eds., 1991, *Principles of Neural Science*, 3d ed. (New York: Elsevier), 486.

103: Corbis-Bettmann.

104 bottom: © Ted Levin/Animals Animals.

105: Adapted from T. Kemp, 1982, *Mammal-like Reptiles and the Origin of Mammals* (London: Academic Press), 312.

107: O. Louis Mazzatenta/National Geographic Image Collection.

108 top: From P. Ramón y Cajal, 1922, El cerebro do los bactracios, *Libro en Honor de D. S. Ramón y Cajal*, vol. 1 (Madrid: Publicaciones de la Junta para el Homenaje a Cajal), 13–59.

112: Lennart Nilsson/Albert Bonniers Förlag AB; Lennart Nilsson, 1990, *A Child Is Born* (New York: Dell Publishing).

113: A. Smith Fernandez, C. Pieau, J. Repérant, E. Boncinelli, and M. Wassel, 1998, Expression of the *Emx-1* and *Dlx-1* homeobox genes define three molecularly distinct domains in the telencephalon of mouse, chick, turtle, and frog embryos: implications for the evolution of telencephalic subdivisions in amniotes, *Development* 125:2099–2111. Used with permission of the Company of Biologists Ltd.

115: Adapted from J. Allman, 1990, Evolution of neocortex, *Cerebral Cortex*, vol. 8A, ed. E. Jones and A. Peters (New York: Plenum Press), 269–284, and J. Pettigrew, Binocular visual processing of the owl's telencephalon, *Proceedings of the Royal Society of London*, B, 204: 435–454.

117: © Don Enger/Animals Animals.

118: © Mitsuaki Iwago/Minden Pictures.

119: Y. Yoshihara, S. Oka, Y. Nemoto, Y. Watanabe, S. Nagata, H. Kagamiyama, and K. Mori, 1994, An ICAM-related neuronal glycoprotein, telencephalin, with brain segment-specific expression, *Neuron* 12:541—553.

Chapter 6

Opening image: By permission of the British Library.

123 right: © Robert Maier/Animals Animals.

124 left: © Gerald Cubitt.

129: Adapted from J. Allman, 1977, Evolution of the visual system in the early primates, *Progress in Psychobiology and Physiological Psychology*, ed. J. Sprague and A. Epstein (New York: Academic Press), 1–53.

130: Adapted from P. Schiller and M. Stryker, 1972, Single-unit recording and stimulation in superior colliculus of the alert rhesus monkey, *Journal of Neurophysiology* 35:915-924.

131: Adapted from J. Kaas, R. Guillery, and J. Allman, 1972, Some principles of organization in the dorsal lateral geniculate nucleus, *Brain, Behavior and Evolution* 6:253–299.

132: left, the author's histology collection; right, adapted from S. Petersen, F. Miezin, and J. Allman, 1988, Transient and sustained responses in four extrastriate visual areas of the owl monkey, *Experimental Brain Research* 70:55–60.

133: B. Dubuc and S. Zucker, 1995, Indexing visual representations through the complexity map, *Proceedings of the Fifth International Conference on Computer Vision*, Cambridge, Mass.; © 1995 IEEE.

134: T. Albright, R. Desimon, C. Gross, 1984, Columnar organization of directionally selective cells in visual area MT of the macaque, *Journal of Neurophysiology* 51:16–31.

135: © John Gerlach/Earth Scenes.

136: Adapted from J. Allman, F. Miezin, and E. McGuinness, 1985, Stimulus-specific responses from beyond the classical receptive field: neurophysiological mechanisms for local-global comparisons in visual neurons, *Annual Review of Neuroscience* 8:407–430.

140: Adapted from H. Zeigler and H. Leibowitz, 1957, Apparent size as a function of distance for children and adults, *American Journal of Psychology* 70:106–109, and L. Harvey and H. Leibowitz, 1967, Effects of exposure duration, cue reduction, and temporary monocularity on size matching at short distances, *Journal of the Optical Society of America* 57:249–25371, Sensory, learned, and cognitive mechanisms of size perception. *Annals of the New York Academy of Sciences* 188:47–62.

141: Corbis-Bettmann.

142, 143: Adapted from A. Dobbins, R. Jeo, J. Fiser, and J. Allman, 1998, Distance modulation of neural activity in the visual cortex, *Science* 281:552–555.

144: Ferdinand Hamburger, Jr., Archives of the Johns Hopkins University.

145: Courtesy of P. Brown.

146: © Ralph Reinhold/Earth Scenes.

147: N. Hadjikhani, A. Liu, A. Dale, P. Cavanagh, and R. Tootell, 1998, Retinopy and color sensitivity in human visual cortical area V8, *Nature Neuroscience* 1:235-241.

150: Adapted from R. Desimone, T. Albright, C. Gross, and C. Bruce, 1984, Stimulus-selective properties of inferor temporal neurons in the macaque, *Journal of Neuroscience* 4:2051–2062.

151: Photomicrograph courtesy of Patrick Hof.

153: Adapted from G. Pelligrino, L. Fadiga, L. Fogassi, V. Gallese, and G. Rizzolatti, 1992, Understanding motor events: a neurophysiological study, *Experimental Brain Research* 91:176–180.

155: J. Allman and J. Kaas, 1974, The organization of the second visual area (V-II) in the owl monkey: a second order transformation of the visual hemifield, *Brain Research* 76:247–265.

Chapter 7

Opening image: Scala/Art Resource, New York.

162: Adapted from R. Martin, 1983, *Human Brain Evolution in an Ecological Context*, 52d James Arthur Lecture on the Evolution of the Human Brain (New York: American Museum of Natural History).

163: L. Aiello, 1992, Body size and energy requirements, *The Cambridge Encyclopedia of Human Evolution*, ed. S. Jones, R. Martin, and D. Pilbeam (New York: Cambridge University Press), 41.

164: Kenneth Garrett/National Geographic Image Collection.

164, 169 bottom, 172: Adapted from J. Allman, T. McLaughlin, and A. Hakeem, 1993, Brain weight and life span in primate species, *Proceedings of the National Academy of Sciences*, 90:118–122.

166: Courtesy of Wally Welker.

167: © Art Wolfe.

168, 203: Adapted from L. Aiello and P. Wheeler, 1995, The expensive-tissue hypothesis, *Current Anthropology* 36:199–221.

169 top: Raymond T. Bartus, Lederle Laboratories, American Cyanamid Company; cover image, 30 July 1982, *Science* 217, no. 4558.

170: Adapted from M. Hofman and D. Swaab, 1992, The human hypothalamus: comparative morphometry and photoperiodic influences, *The Human Hypothalamus in Health and Disease*, ed. D. Swaab, M. Hofman, M. Mirmiran, R. Ravid, and F. Van Leeuwen (New York: Elsevier), 133–150.

171: Adapted from M. Mirmiran, D. Swaab, J. Kok, M. Hofman, W. Whitting, and W. Van Gool, 1992, Circadian rhythms and the suprachiasmatic nucleus in perinatal development, aging and Alzheimer's disease, *The Human Hypothalamus in Health and Disease*, ed. D. Swaab, M. Hofman, M. Mirmiran, R. Ravid, and F. Van Leeuwen (New York: Elsevier), 151–163.

175: Adapted from M. Holliday, 1971, Metabolic rate and organ size during growth from infancy to maturity, *Pediatrics* 50:590.

176: Data from H. Smith, T. Crumett, and K. Brandt, 1994, Ages of eruption of primate teeth, *Yearbook of Physical Anthropology* 37:177–231, and N. Rowe, 1996, *The Pictorial Guide to the Living Primates* (East Hampton, N.Y.: Pagonias).

177: top, Paul Flechsig, 1920, *Anatomie des menschlichen Gehirns und Rückenmarks auf myelogenetischer Grundlage*, vol. 1 (Leipzig: Georg Thieme), plate X, figure 4; bottom, by permission of the British Library.

178 top, 180, 181 bottom, 183, 184 (table), 186: Adapted from J. Allman, A. Rosin, R. Kumar, and A. Hasenstaub,

1998, Parenting and survival in anthropoid primates: caretakers live longer, *Proceedings of the National Academy of Sciences* 95:6866–6869.

178 bottom: © Hugo van Lawick.

179: © Art Wolfe.

181 top: Adapted from D. Chivers, 1974, The siamang in Malaysia, *Contributions to Primatology* 4:1–335.

182: © Art Wolfe.

184: Data from P. Olesiuk, M. Bigg, and G. Ellis, 1988, Life history and population dynamics of resident killer whales (*Orcinus orca*) in the coastal waters of British Columbia and Washington state, *Report of the International Whaling Commission* 12:209–243.

185: top, Dr. Sam Ridgway, U.S. Navy Marine Mammal Program; bottom, © 1994 Brandon Cole/Mo Yung Productions.

188: photomicrographs, courtesy Dr. Hideo Uno, Wisconsin Regional Primate Center; graph, from H. Uno, R. Tarara, J. Else, M. Suleman, and R. Sapolsky, 1989, Hippocampal damage associated with prolonged and fatal stress in primates, *Journal of Neuroscience* 9: 1705–1711.

190: Adapted from P. Garber, 1997, One for all and breeding for one: cooperation and competition as a tamarin reproductive strategy, *Evolutionary Anthropology* 5:187–199.

191: Nick Gordon/Survival Anglia Ltd.

194: Table derived from C. Ruff, E. Trinkaus, and T. Holliday, 1997, Body mass and encephalization in Pleistocene *Homo*, *Nature* 387:173–176.

195: Adapted from T. Crowley and G. North, 1991, *Paleoclimatology* (New York: Oxford University Press), 112.

197: From R. Lewontin, 1978, Adaptation, *Scientific American* 239:212–230.

202: From B. Jacobs, M. Schall, and A. Scheibel, 1993, A quantitative dendritic analysis of Wernicke's area in humans, II, Gender, hemispheric, and environmental factors, *Journal of Comparative Neurology* 327:97–111.

204: top, © Peter Weimann/Animals Animals; bottom, adapted from D. LeBoeuf, 1996, *The Wolf* (Buffalo, N.Y.: Firefly).

205: *White Wolf*, National Geographic Video, 1988.

# INDEX

## Selected Books in the Scientific American Library Series

MOLECULES AND MENTAL ILLNESS
by Samuel H. Barondes

EARTHQUAKES AND GEOLOGICAL
DISCOVERY
by Bruce A. Bolt

CONSERVATION AND BIODIVERSITY
by Andrew P. Dobson

LIFE AT SMALL SCALE: THE BEHAVIOR
OF MICROBES
by David B. Dusenbery

THE ANIMAL MIND
by James L. Gould and Carol Grant Gould

ATMOSPHERE, CLIMATE, AND CHANGE
by Thomas E. Graedel and Paul J. Crutzen

CONSCIOUSNESS
by J. Allan Hobson

EYE, BRAIN, AND VISION
by David H. Hubel

COSMIC CLOUDS: BIRTH, DEATH,
AND RECYCLING IN THE GALAXY
by James B. Kaler

VIRUSES
by Arnold J. Levine

THE ORIGIN OF MODERN HUMANS
by Roger Lewin

PATTERNS IN EVOLUTION: THE NEW
MOLECULAR VIEW
by Roger Lewin

HUMAN DIVERSITY
by Richard Lewontin

POWERS OF TEN
by Philip and Phylis Morrison and the Office
of Charles and Ray Eames

IMAGES OF MIND
by Michael I. Posner and Marcus E. Raichle

AGING: A NATURAL HISTORY
by Robert E. Ricklefs and Caleb Finch

CYCLES OF LIFE: CIVILIZATION AND
THE BIOSPHERE
by Vaclav Smil

THE EMERGENCE OF AGRICULTURE
by Bruce D. Smith

DRUGS AND THE BRAIN
by Solomon H. Snyder

INVESTIGATING DISEASE PATTERNS:
THE SCIENCE OF EPIDEMIOLOGY
by Paul D. Stolley and Tamar Lasky

DIVERSITY AND THE TROPICAL RAINFOREST
by John Terborgh

GENES AND THE BIOLOGY OF CANCER
by Harold Varmus and Robert A. Weinberg

If you would like to purchase additional volumes in the
Scientific American Library, please send your order to:

Scientific American Library
41 Madison Avenue
New York, N.Y. 10010